ECOSYSTEMS
AND
RESOURCE CHOICES
Lessons from the Kerala Floods

ECOSYSTEMS AND RESOURCE CHOICES

Lessons from the Kerala Floods

Editors

R.K. Suresh Kumar

M.A. Ph.D

P. Suresh Kumar

M.A. LL.B., Ph.D

P. Sukumaran Nair

M.A. Ph.D

C. Achutha Menon Foundation

Thiruvananthapuram

KALPAZ

ISBN: 978-93-5324-947-2

First Published, 2022

Published by

Kalpaz Publications
C-30, Satyawati Nagar,
Delhi – 110052
Ph.: 011-47034999, 9811692060
www.kalpazpublications.com
E-mail: kalpaz@hotmail.com

Printed at: Gyan Press, Delhi.

Dedicated to the fond memory of
Prof. (Dr) K. Raman Pillai,
a socially conscious academic of highest integrity,
who led C. Achutha Menon Foundation for over a decade.

Contents

Foreword

The 2018 floods in Kerala has compelled us to reopen the debate on our 'Resource Choices' taking into consideration our ecosystems. The floods affected the livelihood of millions and caused huge damage to physical and natural infrastructure of the state. However, it has also instilled in the people a sense of confidence, revived their urge for collectivism and poised them to think for a stable future. The public domain in the state presently delves extensively on various plans and strategies to rebuild Kerala. The people of Kerala have realized that the ecosystem of the state is going to be central in determining their living conditions. However, it should be understood that for many, this is not a new realization that has occurred in the post flood scenario. For the last two decades the C. Achutha Menon Foundation has been seriously engaged in conducting research on Kerala economy, polity and society. The number of seminars held and the books published by the Foundation is a standing testimony to this.

The narrative on Kerala from such studies will provide an appropriate context to understand the importance and relevance of a book like the present one. Basically, the underlying framework of Governance in Kerala is provided by the ideological conceptions of the communist party and the left forces on the one hand and the call for actions and activities by reformers and social leaders. On the basis of socio, economic and political principles of equality, the governance of the state has succeeded in ensuring a fair and democratic access to all its people over the resources of the state. This approach has empowered the people with a fair level of education, health, dwellings and living sustenance though it has not reached to the level of affluence. Even today there is wide acclaim for these efforts as many hail it as the "Kerala Model of Development".

Later, a paradigm shift took place in the governance of the State under the globalization measures breaking the evolved socio-economic ethos of the state. The two remarkable characteristics of the state, 'equality' and 'collectivism' were replaced by the neo-liberal policies and strategies.

Untamed market principles, unaccounted capital surge and its inflows found a free play. This neo-liberal phase has distorted the pattern of growth of the economy. This has deprived the public of their jobs, skills and living sustenance and left them vulnerable to market vagaries. It is not the market mechanism perse that is to be criticized. The unregulated neo-liberal policies misdirected the growth structure of the economy. Investment activities were channeled to earn abnormal profit exploiting the structural changes in the demand pattern of the people. Though growth rate picked up in the state, the structure of the growth was not conducive for the making of a stable society. It exerted excessive pressure on the available resources often wreaking the ecosystem of the state. The plundering of natural resources continued and no serious efforts were made to address the imbalance.

To uphold the values of 'equality' and 'collectivism' and at the same time to speed up the development of the state, principles of co-operativism and strengthening of public ownership were attempted by the Left governments. However, these strategies were not enough to offset the capitalist grip over the resources in the economy. The misdirected growth and extensive marginalization and inequality of the people prompted a search for an alternative development to the existing market driven governance of the state.

It was felt that a holistic development path based on the principle of sustainable growth would be an appropriate strategy. The Achutha Menon Foundation had organized a two day seminar in 2016 on the theme of "Sustainable Development" and worked out strategies for building an alternative development path for Kerala rooted on the principles of equality and collectivism as experienced from the Kerala Model Development (KMD). It may be noted that the Left Democratic Front (LDF) government in Kerala has initiated various aspects of sustainable development in its policy of 'Nava Kerala'. It is expected that by incorporating the basic principle of sustainable development in its governance, Kerala will succeed in revitalizing the different socio-economic activities in such a way that our resource base will cater to the needs of all people within the collective milieu the Keralites are accustomed with. Within a broad strategy of sustainable development, it would be possible to strengthen our democratic institutions to make suitable policy formulations and its subsequent implementation.

In the background of the present catastrophe we cannot wait for a long time frame for policy making and implementation. The problems of the people are right in front of us. We have to find out urgent remedies to many of these problems and at the same time it cannot be patch up measures. This is a serious issue that we have to address immediately at the governance level.

It is heartening to note that many keralites including the Malayalee's abroad have taken all these issues close to their heart. They have provided their ideas for the reconstruction of Kerala along with financial support. Many experts and specialists have suggested appropriate strategies and frameworks for recreating and rebuilding Kerala which are available both in print and electronic media. This book proposes a concrete set of policies based on the stregth of knowledge and understanding about Kerala's ecosystems and resource choices.

The C.Achutha Menon Foundation has conducted a two day seminar on the theme "Common Property and the Citizens Rights : Issues of Reconstruction of Kerala" post 2018 floods. The seminar in particular examined the use of the natural resources of the state for the common good of all. The importance of Common Property and Citizen's Rights is already engrained in the basic framework of governance of the state. However, the neo-liberal practices that was in vogue for a considerable period has seriously eroded its strength in governance. In the present context, the task is how to translate the citizens' rights over the resources on the basis of collectivism. In our efforts to rebuild Kerala on these lines I foresee two pertinent challenges. One is the consumerist culture in Kerala which has strengthened in the state since globalization. The second is the withdrawal of the government from a proactive role succumbing to capitalist pressures. The spirit of cooperation and collectivism that was evident across Kerala during the calamity gives us fresh hope to face the challenges of consumerism of the people and passivism of governance. Moreover, the vitality of collectivism has brought into public domain the responsibility of the government in rebuilding Kerala, as the private capital remains indifferent.

To ensure the rebuilding of Kerala based on the people's aspirations and ecological concerns, some suggestions can be put forth from the experiences of the Kerala Model of development. Firstly, the specification for each sector and each issue has to be worked out in consistence with the eco-sensitive conditions existing in respective spheres.

Secondly, all Keralites are stake holders in this re-building process. It is only through collective effort that the rebuilding process can be successfully completed. Therefore, transparency is a paramount factor in formulating appropriate policies. The policies taken should be people centric.

Finally, the kind of reconstruction that we envisage in various sectors should be based on decentralized decision process by which grass roots democracy can be strengthened ushering in a meaningful control over resources by the general public.

This book I am sure, is going to be a milestone in the efforts to rebuild Kerala by upholding the principles of equality, collectivism and strengthening the democratic process at the grass root level within the wider context of sustainable development and a balanced ecosystem.

Kanam Rajendran
President
C.Achutha Menon Foundation
Thiruvananthapuram

Thiruvananthapuram
01.01.2022

Acknowledgements

We value very much the support and encouragement extended to us by Kanam Rajendran, President of C.Achutha Menon Foundation and its Secretary N.Shanmughom Pillai. We are grateful to the Members of the Executive Committee and staff of C.Achutha Menon Foundation for being supportive throughout. A special word of thanks to Bindu.N for typing the manuscript with utmost dedication. The editors also express their sincere thanks to Gyan Publishing House, New Delhi for bringing out the book in an elegant manner.

Editors

List of Contributors

Prof. (Dr) M.A. Oommen :	Professor Emeritus, Post Doctoral Scholar of the Rockfeller Foundation, Visiting fellow at the Yale University and a Senior Fulbright Scholar.
Prof.(Dr) B. Vivekanandan :	Former Chairman, Centre for American and West European Studies, Jawaharlal Nehru University, New Delhi.
Dr.P. Sanal Mohan :	Professor, School of Social Sciences, Mahatma Gandhi University, Kottayam.
John Kurien :	Visiting Fellow, Azim Premji University, Bengaluru, Former UN/FAO Post-Tsunami Fisheries Co-Management Advisor in Aceh.
Prof. Udaya S. Mishra :	Professor, Centre for Development Studies, Prasanth Nagar, Thiruvanathapuram, India.
Prof. (Dr) Janki Andharia :	Dean-Jamsetji Tata School of Disaster Studies & Chairperson, Centre for Disasters and Development, Tata Institute of Social Science (TISS), Mumbai.
Tania Teresa Mathews :	Research Officer on a Project for the Tata School of Disaster Studies (TISS), Mumbai.
Dr. N. Anil Kumar :	Executive Director, MS Swaminathan Research Foundation (MSSRF), Chennai.
Suma Vishnudas :	Scientist, Community Agrobiodiversity Centre, M.S. Swaminathan Reserach Foundation, Wayanad.
Prof. (Dr) C. P. Rajendran :	Professor, Geo-dynamics Unit, Jawaharlal Nehru Centre for Advanced Scientific Research, Bengaluru.
Prof. (Dr) Kusala Rajendran:	Professor, Centre for Earth Science, Indian Institute of Science, Bengaluru.
Dr.N. Vinod Chandra Menon:	Founder Member, National Disaster Management Authority (NDMA), Government of India & Senior Advisor, Sphere India.
Prof. (Dr) N.C.Narayanan :	Professor, Indian Institute of Technology, Bombay.
Prof. (Dr) S. Janaka Rajan :	Former Professor & Director, Madras Institute of Development Studies (MIDS), Chennai. President, South Asia Consortium for Interdisciplinary water Resources Studies, Hyderabad.

Dr.V Raman Kutty	:	Professor, Sri Chithra Institutie of Medical Science and Technology, Thiruvananthapuram.
Dr.V.S. Vijayan	:	Former Chairman, Bio-diversity Board and Dr. Sálim Ali Foundation. Member, Gadgil Committee.
Saji John	:	Former Director and Chief Executive Offier, Vegetable and Fruit Promotion Council, Keralam.
G.Madhusoodanan I.A.S.	:	Founder-Director General of World Institute of Sustainable Energy (WISE), Pune.
Prof. (Dr) Mary George	:	Former Professor & Head, Department of Economics, University College, Thiruvananthapuram & Former Chairperson, Kerala Public Expenditure Review Committee, Government of Kerala.
Dr. Gayathri Balagopal	:	Independent Resercher based in Chennai. She has been a Consultant with Madras Institute of Development Studies, Navajbai Ratan Tata Trust & The Banyan Academy of Leadership in Mental Health.
Dr. M. Suresh Babu	:	Professor, Dept. of Humanities and Social Sciences, IIT Madras, Chennai.
Dr. K Jafar	:	Associate Professor, Madras Institute of Development Studies (MIDS), Chennai.
Dr. K. Anilkumar	:	Professor, College of Veterinary and Animal Sciences, Kerala, Veterinary and Animal Sciences University, Mannuthy.
Dr. P.K. Manoj	:	Professor, Department of Applied Economics, (CUSAT), Kochi, Kerala.
Prof. (Dr) Nimmi Kurian	:	Professor, Centre for Policy Research, New Delhi, & Member, Editorial Advisory Board, *Social Science in Perspective.*
Shekhar Pathak	:	Taught in Kumaon University, Nainital and now associated with PAHAR and edits the journal *Pahar*.

Introduction

The well acclaimed development experience of Kerala is popularly known as Kerala Model of Development (KMD). People across the spectrum interpret Kerala Model of Development differently based on the way they look at it. The term has been used extensively in the development discussions related to Kerala since 1970s mainly to indicate its achievements in social sectors such as health, education, land reforms, public distribution system, social security systems and so on. Its accomplishments in literacy and life expectancy rates, fair access to healthcare and low infant mortality and death rates are fairly comparable with that of many of the developed economies. What made this a unique model is the fact that all these achievements were realized with a low level of per capita income. What has caused such an unusual development experience remains a puzzle in the context of development narratives of other states or even countries. This phenomenon has always intrigued political and economic thinkers. Debates among experts has led to various explanations ranging from historical, sociological, economic, political and to religious factors as reasons for the sprouting of Kerala Model of Development. The popular assertion is that a high level of political participation and activism among ordinary people along with a left leadership has strengthened the Kerala Model of Development. It has succeeded to entrench into and transform the state's society, economy and polity. It became a central pillar around which the subsequent governance of the state took place. It has provided a distribution mechanism, structurally calibrated to take care of the basic necessities and ensuring access to material resources. The Kerala Model of Development, however has many inbuilt limitations and it faces serious challenges particularly in the context of globalization.

The relevance of the model was questioned mainly for its incapacity to speed up the rate of growth considering the remarkable spread of the forces of globalization across almost all countries including India especially after 1990.Under globalization, securing higher growth has become a dominant objective of economies. The performance and success of the economies were measured in terms of growth rates in GDP. In India also market forces were unleashed under a liberal frame and subsequently the national economy started leveraging from its inertia showing unprecedented pickups in its growth rate. Several Indian states too reorganized their production process on the basis of the new market rules and principles. As against these unfolding economic scenarios, the situation in Kerala was quite different. Its growth rate was dismal and the basic sectors such as agriculture and industry were lagging without any triggering power to speed up growth. Moreover, the continuous fiscal and revenue deficits prevented the economy to bootstrap its growth potential. Even the sustainability of the Kerala Model of Development was doubted and questioned by many because of the persistent financial conundrums. Much pressure was exerted on the governance of the state of Kerala to streamline the policy framework in terms of new market rules of globalization. Continuous fiscal problems, lower growth rate, dismal performance of major sectors of the economy and the internal urge for reforms necessitated new strategies and upshots in the governance of the state. It was a difficult and complex choice. In the context of an upticking of robust growth rate due to economic reforms at the national level and also in other states, it became hard to discard market led reforms in the state. At the same time, any acceptance and implementation of market reforms is taken to be in conflict with the general characteristics and ideology of the Kerala Model of Development. Thus, unlike anywhere in India, the Kerala government was compelled to adopt a dual policy in its governance, providing a contested terrain both in its economic decision making and later in its political arena.

In this context, it is worth to look into the outcome of the reform measures implemented in the state in the back ground of Kerala Model of Development. The reform measures undertaken in the state has accelerated the growth rate of the economy. It has also caused commendable structural changes in the economic scenario of the state. A visible change in the structural character of the economy was such that the tertiary or the service sector has transformed to make the highest share of the state domestic product followed by industrial and agriculture sector. It is pertinent to note that the reform measures appreciably raised the growth rate. However, the dualistic nature of the governance impelled both by the specificities of the Kerala Model of Development and the forces of globalization has produced a complex outcome in the socio-economic fabric of the state. In the economic sphere, though market measures made a positive turn around in the rate of the state domestic product, it has upset the traditional balance among the basic sectors in the state. In growth process, it is true that, the structure and pattern of the economy undergoes tremendous changes. The historic transformation pattern is such that the service sector takes a lead over and above the two other basic sectors viz the agriculture and industrial sectors. In this process of structural changes, a healthy balancing between the sectors will emerge triggering a composite growth of the state domestic product. However, in the state of Kerala, under the liberalization policies not only the basic balance between the sectors has been lost but also the industrial and agriculture sectors started shrinking relatively to the services. Even in the services what has occurred was not an even development. The organic link among the leading sectors in Kerala was lost due to the complexity or dualism that prevailed in the governance of the state. For example, the Kerala Model of Development has created a demand structure for collectives and public goods such as demand for education and health facilities and other public services. The KMD has orchestrated a collective demand pattern which was difficult to be sufficiently satisfied within the framework of a welfare-oriented programme due to constraints in state finances.

It is the foreign remittances particularly from the Malayali Gulf migrants that have provided a strategic turning point in uplifting the Kerala economy from a low-level equilibrium trap. Foreign remittances as an inflow became an important source of income for many of the Kerala households. It became a powerful source of consumerism which subsequently produced a consumerist culture in the state. In fact, it is this consumerist culture that has strongly influenced the kind and pattern of investment that was in vogue during the post reform period in Kerala. In the service sector, much of the investment mainly went into constructions, hoteliering, trade, tourism, education and banking. These investments are made mainly by the local capitalists for whom investment opportunities were neither available nor encouraged under KMD. Generally, under KMD, the entrepreneurial activities by private individuals were not encouraged. It was only after the release of free market forces that the hither to lazy and hidden capital surfaced and flowed to safe and quick profit yielding and rent seeking areas of investment.

The forces of globalization, thus finally did raise the growth rate of the Kerala economy. But unlike in other states, this upticking in growth has not helped the state to spiral an independent organic growth. Rather, the sectoral imbalances and malinvestments has produced many social and economic issues which the state has never faced before. All such issues that cropped up are, in fact, a challenge to KMD and undermined the credibility of it. For example, particularly in the higher education scenario, the post reform period has witnessed mushrooming of many self-financing institutions in engineering, medical and paramedical areas and also in arts and science subjects feeding upon the appetite of the people for better education and employment opportunities. This hunger for better future, induced and pushed many of the households to take bank loans and find other sources of finances including drawing upon of their savings and disposal of their physical holdings to meet the higher expenses required for the new educational opportunities. The whole matter worked like a network where the beneficiaries are the banks and other financiers on the one hand and the new capitalists who found an opportunity to reap profit from the kind of socio-economic structure that evolved in the state. However, it is now clear that such a model has not worked for the benefit of Kerala in the sense that the man power requirement of the state has not bloomed enough to absorb the newly educated graduates in industry, agriculture or in the newly emerging areas of employment. What has been opened for the new entrants in labor markets is petty job opportunities with low wages such as sales girls and boys, drivers and other odd jobs without any kind of job or income security. The fact is that Kerala still continues to remain as a reservoir of the highest number of educated unemployed in the whole country.

It is also a fact that while the educated unemployment is one of the highest in the state, it also accommodates about 50 lakh of migrant workers in different economic activities in the state. It is a strange situation which cannot be explained by conventional logic of economics. A possible situation that justifies this paradox may be that the educated prefer only white-collar job, however poor the wages are. It seems that the KMD has succeeded to ensure a socio-economic security that dares the educated unemployed youth to prefer either a white-collar job with poor wages or remain on unemployment rolls rather than entering as casual workers and daily wage earners. In short, the legacy of KMD in the state of Kerala has neither fructified in to a self-generating auto centric growth by itself, nor appropriated the potential of free market forces to stimulate a balanced multi-sectoral growth. In the governance of the state, the ethos of KMD on the one hand and later the free market forces on the other, has culminated in to a socio-economic structure characterized by development imbalances, growing inequalities, and low paid jobs for the educated labor along with a polarized political and ideological rhetoric.

The concept of development itself has transformed globally in the context of the environmental catastrophe that occurred across the world. Climate change, resource constraints and destruction, over use and over exploitation, bio-diversity loss, rising social inequalities and poverty experienced all over the world have raised questions on the existing development models. Doubts about the sustainability of the existing production systems came into the fore. Ever increasing environmental issues and disasters also prompted the development scientists to give focus to climate change as a paramount issue. An international framework involving economic, political and cultural interdependencies related to development issues became a broad canvas which individual economies have to reckon with. Economies have to accept such frameworks as a regulatory mechanism in making their own development policies. Accordingly, concepts such as sustainable development, equitable cooperation, clean environment and clean technology are embedded not only as instrumentalities in policy formulations but also as globally accepted human rights adherent.

However, historically, the KMD has taken a silent perspective on the environmental issues facing the state. In fact, the land reform measures adopted in the state did not have any statutory provisions to curb the extensive deforestation that took place in the state for the spread of agriculture and unfazed constructions in the eco-sensitive areas. The state has been perceived as an ecologically fragile and sensitive zone by many experts and committees long back. It has been pointed out by many that being a land scarce area, Kerala is not a suitable place for large scale industrialization like its neighboring states. With the publication of Gadgil Committee Report, the eco sensitivity of the state was widely recognized, but nothing concrete has been done for the effective implementation of the recommendations. The socio-economic fabric of the state was such that no political decision could be taken for its implementation though all the political parties subscribed to a common set of values. It is during the three unprecedented floods in 2018, 2019 and 2021 that people of Kerala realized the eco-sensitive reality of their habitats and the environment. It is expected that the post-flood scenario would be a turning point for the people of the state in terms of their livelihood and development concerns. Sustainable development which is a new dimension in the development policies and initiatives at the global level, found its reverberations in many discussions and policy suggestions by leaderships at various levels in the state.

The pressure exerted by the private sector on the natural resources of the state was abnormal especially after the globalization phase. In the absence of any effective guidelines and regulatory mechanisms, the nascent capitalists and their mafiosi has plundered the commons. The growth of the construction sector has wrecked ecological balance through mining and quarrying activities. The much-hyped sustainability was thus unrealizable in the state mainly due to the crony capitalist behavior of the ruling class in complicity with the mafiosi groups operating in the state. The KMD failed in evolving a policy frame to ensure an equitable access to common resources of the state.

Several attempts were made to retain the relevance of the KMD and to relocate it in the context of the misguided and lopsided development since the freeing of the market forces in the state. Decentralized planning, cooperative framework and special purpose platforms are mainly the modes used to strengthen the KMD. The attempt was to contextualize the KMD legacy as an alternative to globalization and to face issues related to climate change and sustainable development. Rather than considering it as a substitute to market-based development policies, the lessons and experiences involved in KMD has to be properly and effectively appropriated, linked and nurtured with market principles. The 73rd and 74th Amendment to the Constitution of India has brought decentralization as a mechanism to broaden and strengthen various dimensions of democracy at the grassroot level to improve the welfare and living conditions of the citizens. The globalization measures have widened the scope of decentralization to appropriate the market forces consistent with local situations and local interests. Wider participation by the people in the decision-making process enable and equip them to put the local economic resources to optimum use in a sustainable manner. Cooperation has also to be used as a tool to strengthen democracy and decentralized decision making particularly in the use and allocation of economic resources. However, the decentralization and cooperative movements as likely forces for reorganization of the socio-economic fabric of the state has not succeeded as it could have. Rival political formations used it as a camouflage mainly to make political mileage and protect their sectarian self-interests. Though the decentralization and cooperative movements are efficient strategic entities at micro level with immense grassroot benefits in resource allocation, poverty alleviation and in protection of livelihood activities of the marginalised sections, its scope to streamline the governance of the state on such principles still remains an overarching ambition. An overwhelming example in Kerala to show how productive and successful it can be at micro level is the 'women empowerment' secured in the state through the Kudumbasree and other programmes. But no industrial or agricultural reorganizations took place under cooperative or decentralized movements in Kerala on a large scale as has taken place in other states like the famous Anand Model in Gujarat.

The process of development of Kerala, as it stands in the context of post flood scenario, is an opportunity to make a critical assessment with a view to strengthen its future prospects. Two questions are important in this regard. Is the KMD legacy sufficient for us to move forward in the development needs of the state in the context of the concept of development transpiring towards a global standard wherein socio-economic and a cultural interface are increasingly taking place? Is it possible for a market driven development structure to take care of the value systems and aspirations that the Kerala society has nurtured through struggles and public actions over a long period of time? The process of development that has taken roots in the state, raised serious questions on the KMD's continued relevance for the future. A conscious effort to shed the drag associated with the KMD in the policy making realms of the state is necessary. Today it is increasingly realized that KMD had its genesis in the past. For example, the contributions made by the erstwhile Travancore rulers to education, particularly to women education; public health, public infrastructure and other public utilities has provided a collective base which became a strong pillar of Kerala Model. The religious leaders and social reformers had sensitivised the people of the values of equality and brotherhood. This base was further emboldened by the public action undertaken by the masses under the leadership of many social and political visionaries. It is on such a fertile soil that the left has planted their progressive slogans particularly relating to economic ideas and demands such as land for the tiller, rights for the workers, minimum wages and a thrust on welfare actions. No formations of any kind could stand indifferent to it. It is in this unique context that the KMD emerged.

Unfortunately, the KMD became a fixity from which it could not free itself objectively. While the KMD can remain as a guiding principle in the governance of the state, for many, it is of rhetoric value only. The Kerala Model of Development mentioned in a study by Prof. K.N.Raj, became convenient for many, to use it to dilute reform measures much needed for steering the growth of the state. Global reforms were found inconsistent and remained as an anathema vis -a -vis the KMD. Moreover, the ideological compulsions of the left forced them to justify the KMD as a development model ever sensitive to the development requirements of the state. Their enthusiasm to defend it as a model capable to be placed strategically as an alternative to globalization further dented the utility of KMD. An objective approach to KMD is required to safeguard the socio-economic interest of the state. Accepting the historical legacy of KMD and freeing it from the assertion that it is a left contribution and monopoly are inevitable for the progressive left to make it more useful and realistic for the contemporary times.

The immediate concern in the context of the limitations of KMD, is how to steer the course of development of the state? In addressing this issue, the first consideration is how to secure sustainable development in the context of the new framework of climate change? How can we make development faster in the state by imbibing a collective frame involved in KMD? How can we chart out a development course that can effectively address the needs and aspirations of the people of Kerala along with the responsibility of safeguarding our ecosystem?

These development issues are the focus in this work. After the ecological disasters in Kerala in 2018 and 2019 it was felt by the Achutha Menon Foundation that the state cannot continue its present development strategies unabated for long. The KMD has to be re-evaluated in the context of pressing development needs of the state. Also, the development policies and strategies can be streamlined only by following national and global guidelines. Keeping these parameters as a broad frame, the Foundation had organized a two day national seminar in which statesmen, experts, academicians and researchers from all over India participated for framing a working model for the construction of Post Flood Kerala.

The Seminar which was held on 25th and 26th of October 2018 discussed the theme 'Common Property and Citizens Rights: Issues of Reconstruction of Kerala'. The theme was chosen primarily for two reasons. The Foundation, in an earlier research study had found the necessity for incorporating sustainable development policies in Kerala's development process. Though conceptually the theme has been recognized in policy frame, no effective steps were implemented to transform the development process linking it with a sustainable ecosystem. The post flood scenario in Kerala has reinforced the necessity for its adoption in the growth process. Secondly, though the KMD is beset with serious challenges in its development outcomes, there is a near unanimous understanding that equity, collectivity and social welfare mechanisms have to remain as the central objectives of the governance of the state. The theme thus examines the mechanisms by which a sustainable growth process can be initiated consistent with the inherent value systems embedded in KMD in the context of post-flood scenario in Kerala. A total of 23 papers are selected for publication in this volume. These papers are arranged under three subtitles reflecting different dimensions involved under the comprehensive theme titled as *Ecosystems and Resource Choices: Lessons from the Kerala Floods.*

1. Resource Base and Common Property Rights: Retaining the Legacy of KMD

Keeping the popular ethos of the KMD, a rebuilding process of Kerala economy requires a concrete assessment about the quantum of resource availability and its equitable and efficient access to its people. Here our approach to governance assumes importance. Under this sub-title, seven papers are included which by and large, indicate the kind of policy frames that are to be formulated consistent with the legacy of KMD. M.A.Oommen, in his article, *On the Threshold of a Grand Opportunity: Salience of an Approach to Rebuilding Kerala* stresses on the necessity of long term policy choices which are to be bold and dynamic. He wants to make the local governments the epicenter of recovery and rebuilding. B.Vivekanandan in his article suggests that the 'The New Kerala' should be a modern welfare state. He reminds the Swedish experience and hopes that a dynamic, statesmanly political leadership would invariably endeavour to make catastrophic situations into new opportunities to pave the way for augmenting the general welfare of the whole population. Taking a critical look at the KMD, Sanal Mohan in his article *Citizenship and Reimagination of Keralam* raises questions on citizenship particularly probing the experiences of the marginalized sections of Keralam. John Kurien in his paper *Rebuilding Kerala: Lessons from Aceh* highlights the need for a renewed political engagement in a new framework to create a new Kerala. While discussing on the *Social Protection in Post-disaster situation in Kerala,* Udaya S. Mishra argues for a reorientation of social protection programs taking into consideration the core development and livelihood issues of the state. The paper authored by Janki Andharia and Tania Teresa Mathews focuses on the importance of environmental justice and the governance of Common Property Resources which together work towards environmental sustainability and engendering community participation. Empowering local self-government to maintain bio-diversity and to secure sustainable development is the central theme proposed jointly by N.Anil Kumar and Suma Vishnudas in their article.

2. Rebuilding Kerala:Post Flood Challenges

Under the second title, the growth process that has to be unfolded in Kerala economy in the context of post flood scenario is addressed. This section consists of fourteen articles covering the major sectors and analyzing thread bare the priorities and strategies that are to be framed to streamline growth in Kerala. There are five articles which are directly focusing on the different aspects of flood related issues. C.P. Rajendran and Kusala Rajendran in their article *Post Flood Vision of a New Kerala* suggests alternate de-growth options for the Post Flood reconstruction of a new Kerala.

N. Vinod Chandra Menon provides an overview of the impact of floods, damage and the loss. The paper also examines the strategies of the government of Kerala with regard to the post flood rehabilitation, reconstruction and recovery. In his article *August 2018 Floods and Landslides in Kerala: From Emergency Response to Reconstruction and Recovery*, Menon emphasizes on the urgency of discarding the lopsided development strategies and the inevitability of eco-friendly sustainable development pathways. Focusing on the development strategy in a local area, Kuttanad, N.C. Narayanan narrates the development interventions that deepened the unavoidable disaster of 2018 Kerala floods. The development interventions in Kuttanad in the past two centuries to make it a rice-centric economy has led to many hydraulic changes that accentuated the situation when floods hit Kerala in 2018.The paper examines the deeper triggers of floods based on historical evolution of land use. The paper highlights the need for a more integrated approach for managing the heterogenous land and water system in the state. *Understanding Disasters and Resilience: Towards Rebuilding Kerala* is an article which details different causative factors that have culminated in frequent floods in Kerala. In the article, the author, S.Janakarajan prescribes pathways for way forward wherein he assigns a paramount role for a government as the protector of ecology and environment.

How floods impact health? Usually, health issues are analyzed in the context of a normal state of affairs in an economy. A pandemic like situation as we face today definitely warrants extraordinary interventions. V.Ramankutty, in his article *Recovery and Reconstruction after the Flood: Challenges from Health Sector* links flood with a new dimension in health sector. He argues that a conventional reckoning of loss of life during floods is missing the whole story. The loss due to death lingers beyond the post-flood reliefs that are offered by the state. Health sector should broaden its services in the post-flood scenario at a greater level than it is presently tuned with.

There are three important papers that farsightedly address the ecological issues pertinent for the people of Kerala in their attempt to rebuild Kerala. The noted ecologist and activist V.S. Vijayan in his paper *Rebuilding of Kerala through Ecological Civilization* presents some vital strategies and points for the sustainable use of resources of the western ghats and the wetlands in the state. Saji John is of the view that climate change will remain as a specific setting in the post flood times and efforts should be made to adopt climate smart agriculture taking into account local producer's knowledge, requirements and priorities. In the article, *Enabling Environment for Climate Smart Agriculture* he stresses the importance of farmer friendly policies. *A Green Energy Future for Kerala* by G.Madhusoodanan is an attempt to project the energy requirements of the state. His modelling shows that by 2050 it is possible to meet 100 % of the state's energy demand with renewable sources.

Specific sectoral growth prospects are analyzed by researchers in the context of post-flood recovery plans in the state. Mary George discusses extensively the issue of sustainable agriculture in her article *Sustainable Agriculture:Past Lapses and the Way Ahead.* A new dimension has been added by Gayathri Balagopal in the rehabilitation and rebuilding attempts through her article by focusing on the need to integrate animal and human health. The articles by M.Suresh Babu, K.Jafar, K.AnilKumar and Manoj P.K deal with effects of floods on employment and livelihood issues, migration, and tourism.

3. Learning Lessons and Experiences

The third part in this volume deals with the national and international experiences regarding flood control and post-flood management. Drawing on examples from Asia, Africa and Europe Nimmi Kurian points out the need for moving beyond policy silos to explore the potential role that non-structural risk reduction can play in building flood resilience. In her article *Of Flows and Flawed Frames: What Can We Learn (and unlearn) from the International Experience on Floods* she argues that flood risk governance has to be acknowledged as a cross cutting concept that cannot be seen in isolation from critical issues of poverty reduction, land-use management, water supply, solid waste management and disaster prevention. Shekhar Pathak in his article *Understanding Kerala (2018) and Himalayan Floods (2013-14)* submits that though the physical conditions in these two regions are different, it is the common modern capitalistic life pattern that has caused the flood fury in both regions. He also dealt with Uttarakhand floods of 2013 and the Kashmir floods of 2014. He observes that the fervour for conservation and equity, understanding the climate changes and their impact on everyday life should be relooked to ensure sustainability and eco-sensibility.

We hope that this volume will also, like the previous ones, be well received by the academic community, political leadership and administrators who are genuinely interested in the development process of Kerala. Through this volume, the C. Achutha Menon Foundation is fulfilling its societal obligation to learn, understand and disseminate the different dimensions of Kerala's development essential for the 21st century growth of the State and its people. Finally, a word of thanks to all the scholars who have promptly responded to our request, invitation and time schedules and to the President of the Foundation Kanam Rajendran who presented, on behalf of the Foundation, 'The Policy Recommendations' to the Hon. Chief Minister of Kerala.

Editors

PART - I

RESOURCE BASE AND COMMON PROPERTY RIGHTS: RETAINING THE LEGACY OF KMD

1

On the Threshold of a Grand Opportunity: Salience of an Approach to Rebuilding Kerala

M.A. Oommen

Rebuilding is different from repairing. This paper outlines the contours of an approach to the rebuilding project underway in Kerala which was caused by a wave of devastating floods and landslides in August 2018 which washed away in a jiffy 4% of the state's GSDP affecting 5.4 million people and displacing 1.4 million while traumatizing 1260 villages and crippling rural life that comprises 52% of its population. We know that this state which is only a little over 1% of India's geographical size has a fascinating human development narrative which has given it a unique place in the development literature. Probably, less well known but equally *sui-generis* is Kerala's rich bio-diversity, one of the most valued resource on this globe, its geographical landscape, accommodating 44 river systems, beautiful lagoons and lakes majestically walled off by the Western Ghats from which the rivers originate and which is home to a rare combination of flora and fauna in this planet. Kerala created history in this world when the Communist Party ascended to power through the ballot box in 1957. The first ministry expanded public sphere, public reason and freedom challenging the worst forms of the caste-class iniquities and oppression. From today's rebuilding launching pad we may recall that the last sixty years saw several social and economic events like the growing inflow of foreign remittances from mid 1970s (in 2018 it was over $15 billion) the heralding of remarkable reforms like abolishing landlordism, universal literacy, affordable health care and ushering in decentralised people's plan for deepening democracy to mention the most prominent. This period coincides with a consumption-induced capitalist paradigm of development that lifted its growth but significantly affected its ecological viability and threw open a regime of commodification of health, education and environment. The ecosystems and resource base of lakes, lagoons, rivers and forests have suffered irreparable loss due to encroachment and indifferent management. No wonder there was social failures like widening inequalities, the marginalization of the most vulnerable sections of society, increase in the crime rate (from 13.3% of all-India in 2014 to 14.7% in 2016), atrocities against women and children, high magnitude of suicide rate, mounting road accidents and so on. The present unprecedented disaster happened when Kerala launched its Thirteenth Five Year Plan and the country chose to dismantle the planning process and embraced a market mediated development. An approach paper was already out before this. Any paper without a vision for Kerala and silent on long-term policy choices could not be counted upon to turn the challenges facing the state into a great opportunity. Kerala needs bold and dynamic policy choices. A green state backed by a green economy that will have the full support of the well-meaning citizens will have to spell out its long-term rebuilding vision, major objectives and priorities besides the strategic options envisaged. The purpose of this paper is to throw some insights that could be useful in developing an approach to rebuild Kerala. This is but indicative and certainly not exhaustive.

That the Government of Kerala produced the Post Disaster Needs Assessment (PDNA) Report a fairly comprehensive document in record time with the support of an able team of experts drawn from eleven UN Agencies, the Asian Development Bank, the World Bank and the European Union Civil Protection and Humanitarian Aid is no mean achievement. This document provides the foundational launching pad for making a sound approach, a powerful strategy and stable policy framework not only to move ahead for disaster risk reduction and recovery but also for building a new Kerala. It is for the government and the people of the state to give it the shape and structure that the situation warrants. The report does not seem to tolerate red-tape, and corruption. Five months have elapsed and no official approach document is available in the public domain for discussion. This brief paper proposes a few guiding principles of approach that are germane to the rebuilding process and then identify just four priority areas.

1. A Venn Diagram Approach Underpinned by Transparency Guarantees

While adversarial politics is part of democratic practice, extremely divisive and polarizing approach will be unhelpful especially in a rebuilding context. I think the state needs a ***Venn diagram approach*** illustrated below for taking the project to successful fruition.

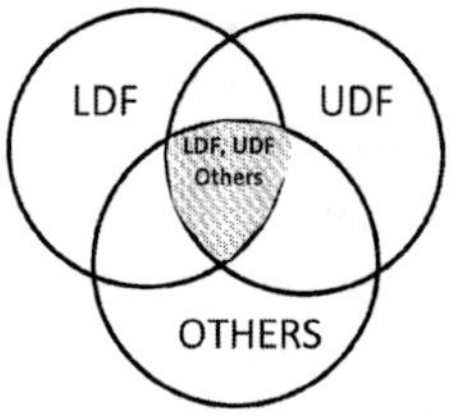

Fig. 1

For facilitating this the most important prerequisite is to ensure some *transparency guarantees*. There is also need to perfect new political idioms through discussions. No society can meaningfully function without some presumption of trust and openness that citizens can expect. Transparency guarantees is an instrumental freedom "in preventing corruption, financial irresponsibility and underhand dealings" [Sen 1999:.40]. In the present context the first step to win the trust of the people including the opposition political parties is to place some transparency guarantees in the public domain. Certainly a dominant rent-seeking politics and the growing disregard for rule of law cannot be the basis for rebuilding Kerala.

A white paper that will spell out the sources and uses of funding along with a frank critique of fiscal management of the past as well as the differentia envisaged under the new dispensation is the need of the hour. We know that the state did not achieve any of the targets fixed in its Medium Term Fiscal Plan. There was heavy tax revenue loss due to tax evasion and avoidance. In the gold sector in 2016-17 alone there was an estimated transaction worth Rs.2924.8 crore that was outside tax network as per the CAG Report. The yawning gap between interest payments and pension payments since 2009-10 is a matter of serious concern for the state government in the present context and must be brought to the notice of the civil servants as well as the public of Kerala. Raising the retirement age to 60 years, suspending the five year pay-revision commission for some time till we come a good way with the rebuilding process have to be re-examined in the present context. Persistent savings in budget allocations and imprudent demand for supplementary grants are signs of poor fiscal management. That the State Disaster Management Authority constituted in 2011, utilized only a little over 43% of the funds in the test-checked districts by the CAG from April 2012 through March 2017 shows the continued inefficiency of this institution which can be tolerated only at our risk. An honest review of fiscal management may be made, from which new reform options should emerge.

The PDNA estimates an expenditure of Rs.31000 crore for five years at the 2018-19 prices. This works out to Rs.6200 crore per annum. If we consider that 35% of this cost will be accounted for by the beneficiaries and the private sector the actual cost will be only Rs.4030 crore per annum at constant prices and allowing a 5% rate of inflation the estimate for 2019-20 will be only Rs.4321 crore. My hunch is that this will be below 3% of the budget for 2019-20. Given Kerala's tax revenue potential to be mobilized through prudent fiscal management and the enthusiastic response to finance the recovery and rebuilding project by the people including the diaspora the goal is not an uphill task. But it is important to clarify the roles of KIIFB (Kerala Infrastructure Investment Fund Board), whose mandate is outside the rebuilding project and the Kerala Bank. The KIIFB apparently with a good technical bearing has approved 469 projects costing Rs.39714 crore till December 2018. It has an ambitious programme of market borrowing. But only three of the projects seem to be self-liquidating. The implications need be debated.

The Chief Minster's Disaster Relief Fund (CMDRF) is heading towards Rs.3,400 crore. If the government employees and some dedicated diaspora contribute to the funds at regular intervals a steady flow of funds could be ensured. The Corporate Social Responsibility (CSR) is being explored and could easily yield Rs.150 crore per annum. Utilization of the National Disaster Response Fund, additional funds via centrally sponsored schemes and several other expected sources have to be spelt out in detail. The scheme of additional taxation notably via the 1% of cess on GST for next two years, the magnitude of borrowing envisaged beyond FRBM ceilings and so on call for a white paper which will spell out in detail the trajectory of funding that is being envisaged. This is the best way to promote accountability, transparency and trust. The rationale of the White Paper is because public money has to be spent with wisdom, faithfulness and efficiency.

2. Recapturing the Renaissance Ethos

A common refrain in recent months widely heard from all vocal sections notably from the ruling and opposition political parties is to recapture the spirit of renaissance in Kerala. This indeed is timely. Lest it becomes a part of the usual political rhetoric one need to clarify what recapturing renaissance means in the present context. I think it is discovering and rediscovering the essence of humanity which has been under a cloud in recent times. We know that the process of renaissance in Kerala gave a culture shock to a community steeped in hierarchy and unfreedoms of the worst order. Like the Enlightenment of Europe that threw up reason against religious fundamentalism, Sree Narayana Guru brought out the significance of freedom for human beings who were deeply enslaved by caste-class iniquities. His fervent plea was to get enlightenment and reasoning through education thereby seeking to strike a mortal blow to the holy/unholy or touchability, untouchability divide. The terms 'Manushya Jyatai' and 'Sadhu Janam' coined by the Guru connote the suffering humanity and no wonder was immediately echoed by Ayyankali and copied by several social reformers like Vakom Moulavi who were working towards freeing their communities.

In the present context, it is important to admit that Kerala has receded if not slipped significantly from the spirit of renaissance. The moot question is whether Kerala will sink preconceived religious prejudices and fundamentalism and widen the public sphere imbued with public reason for the common good? You create new norms and values only through public debates that honour reason and justice. Fundamentalists be they religious or political can never engage in reasoned debates and arrive at sound norms and values.

3. Undergird Your Strong Points and Remedy Your Weak Links

Logically, the best approach is to build on your strong points and strengthen your weak links. By this token *inter-alia* two important strong points to be noted are : (a) to build on local governance and decentralised planning system and (b) to foster and rally the solidarity force and voluntarism displayed during the floods especially like that shown by the fisher folk whose selfless actions (4537 fishermen with 669 boats) saved over 65000 lives something unheard of in the history of crisis management in the world. Among the many weak points we may mention the ecological overkill that considerably disfigured the bio-richness and beauty of Kerala and the social failures in integrating the tribals, fisher folk and other historically marginalized communities in the process of development. As the PDNA Report notes:

> "Focusing on the particular vulnerabilities of the excluded offers an opportunity to establish a human right-based approach to recovery and rehabilitation" [p. 47].

In this short note I may identify four priorities.

4. Identifying Four Priorities

(i) Make the Local Governments the Epicenter of Recovery and Rebuilding

The local government or the third tier of governance in Kerala comprising nearly 22000 elected representatives have the strong legacy of an institutionalized multi-stage planning process that has widened the avenues of people's participation. This tier has been the recipient of over 25% of the state's investible resources and should be natural partners in the recovery process. Although local government system of Kerala certainly has empowered women they still do not control the levers of change which can come out only through transforming gender relations. Will the women wall be the harbinger of change?

The current debates and literature are silent on the challenging issue of urban governance in Kerala. Several problems of Kerala arise out of the failure to address the issue of sprawling rural-urban continuum. Kerala's urban population increased from 4.7 million in 1981 to 15.9 million in 2011, a 3.4 times increase in three decades. Urban Kerala has the highest monthly per capita consumption expenditure (MPCE) in India with accelerating demand for consumer durables and conspicuous consumption. This has added to solid, liquid and electronic wastes which are unbearable by any standard. The rebuilding team should not bypass this. In the floods the panchayats have been in the forefront of rescue and search operations and must be roped into as natural partners in the rebuilding project. Now that the state government had already initiated a second plan campaign and a District plan initiative for all the 14 districts enlisting the cooperation of the Local governments meaningfully is possible, it is not difficult to revive the gram sabha spirit of the past as part of this exercise. The plan funds have to be reprioritized and reworked. I think the tremendous doability and action power of the people can be calibrated best for reconstruction only through the local governments which are pathologically getting compromised to a low level equilibrium working as subordinate agents. They have to be made active partners rather than as subordinate agents in the rebuilding project.

We have to rope in the resilience of the local people and keep their motivation intact lest we fail. Reconstruction of 17000 new houses and repair of 46000 houses, Anganwadis, health centres, schools etc have to be done during the next 5 years. Every day solid waste of the order of 10044 tons are produced in Kerala at the local level. There is no abatement to the growing garbages. Removal of the colossal disaster debris be they beddings, broken furniture, white goods or e-waste is a persistent problem which only the local governments can manage. Those regions deeply affected by floods like Kuttanad[1] and those affected by landslides as in Wayanad and Idukki can be salvaged and developed only through equipping local governments. Similarly the recovery of lakes (e.g. Vembanad) can be done only with the spirited cooperation of the local governments. It is important to recall that the people's commission on Vembanad eco-system recommended the constitution of a joint committee of District Panchayats of Alappuzha and Ernakulam to restore Vembanad and make it useful and livable. The PDNA has quite thoughtfully recommended the creation of 70 housing facilitation centres at the rate of one for two blocks and green technologies centres in every village. To avoid a fragmented approach and facilitate coordination, the manner of dovetailing the Nava Kerala Mission, notably the Harita Kerala Mission that seeks to integrate waste management, organic farming and water resource management the Clean Kerala Mission and the Suchitva Mission into a functional entity is immediately needed. Every panchayat should have the expertise to develop local level recovery planning and development. The massive rebuilding of the expertise and resources is a herculean challenge which KILA (Kerala Institute of Local Administration) and IMG (Indian Institute of Management in Government) must take on hand. But you have to make local governments the hub of the recovery programme and the local government elected representatives made the agents of rebuilding.

Section 55 of the Kerala Municipality Act provides for the creation of a state Development Council presided over by the Chief Minister with the Chief Secretary as Convener and comprises all ministers and leaders of panchayats and municipalities. Both UDF and LDF have atrophied this institution. Kerala is now on the threshold of a historic moment and the State Development Council can be made the umbrella organization that will integrate the Nava Kerala Mission and the local governments to serve as the centre-piece of the rebuilding project.

(ii) Making Kerala a Green State

The major plank of the recovery project is to make Kerala a green state as the PDNA report suggests. Rebuilding a green economy cannot be done without understanding the havoc that has been done to the land-use pattern as well as to its beauty and resilience. Reminiscent of Keats' words 'a thing of beauty is a joy forever', Kerala has to be made an outstanding tourist destination besides a thing of joy for the people of the universe in the years to come. Human beings have to rely on Nature's resources for survival, but aggressive inroads invites its wrath like landslides and floods. It also destroys livelihood and creates dystopia (e.g. Eloor, Alappad). The land use and water management of Kerala whose natural division into coastal land, a special ecological mosaic, midland that has 41 west-flowing rivers to enrich it and a high-land of unique forest system destroyed by the migrants from other parts of the state are facing serious threats. The most striking trend is the sharp decline in the area under paddy from over 35% of the cropped area on the eve of the formation of the state a position maintained upto mid 1970s, which fell to a low 7% during the last three years. Unlike the rice cultivation in other parts of India, in Kerala it is done on a water shed basis which commands the highest environmental value per hectare in the world [Costanza, R *etal* (1997)]. The paddy lands of Kuttanad, Kole, Pokkali, Kaipad, Ela etc., provide unique watersheds. They serve as natural drainage paths for flood waters, and conserve ground water and are essential for the preservation of a rich variety of flora and fauna. Unfortunately these valuable resources are fast disappearing. The water scarcity that Kerala faces today is largely linked to this paddy reclamation. The policy makers of Kerala apparently have been feigning ignorance about these existential threats. Although the Kerala Conservation of Paddy Land and Wet Land Act was passed in 2008, which itself was very late in the day, not much has happened except to ratify all illegal transgressions even after 11 years. We should realise that the quarrying, metal crushing and sand mining lobbies are endangering Kerala's eco-system. A well designed land use policy and a comprehensive mining policy have to be made as part of the project. Instructively the Seventh Report of the Legislative Environment Committee (2014-16) regarding the working of the quarries neither analyses, nor addresses the issues directly or in detail. The manner in which the government and the political parties of this state responded to the Western Ghats Ecology Expert Panel (2011) (popularly called Gadgil Committee) report which is a well substantiated and convincingly argued document to put it mildly was irrational and shameful. Placating vocal voter interests can never be the way to building common good and sustainable development. How can any one interested in the sustainable development of this region ignore the geospatial data base and the well-balanced participatory policy guidelines of this scientific report and rebuild Kerala? The PDNA recommends an integrated water resource management approach to build around the theme such as 'room for river' and 'making space for water' as the basis for addressing flood protection instead of creating fresh dams. This approach followed by comprehensive eco-sensitive land use policy and a new policy for building construction based on disaster resilient construction techninques is the need of the hour. It is important to explicitly note that through the state water policy recognised the need to 'integrate the problems and propects of water resource systems by considering the river basin as the basic unit of planning' nothing has happened. We may have to start from here.

(iii) Social Democracy, Social Justice, and Caring for the Poor

The recovery and reconstruction vision statements must explicitly put social justice as their priority. Why the tribals and for that matter the historically marginalized remain neglected is a question that needs to be raised now with added emphasis. It does not speak highly of Kerala's development that as per 2011 census the tribals of Kerala have the highest percentage of dilapidated and unlivable houses. Indeed the disasters have brought to sharp focus once again the vulnerabilities of the poor in general as well as of the elderly, women, the transgender and the disabled. Given Kerala's ageing phenomenon the three million migrant labour will have to continue to be part of our development for years to come. They will have to be duly provided for. Economists are never tired of establishing the importance of the sum-total of value-added and its growth. I think development becomes really value-added only when the poor are genuinely lifted up into the mainstream of society and the economy. Rebuilding has to be seen as a great opportunity to put an end to social exclusion through structural reforms along with short-term measures that will relieve the impact of the disaster immediately.

Given the dramatic climate changes underway in the planet and Kerala's visible resource mismanagement and policy failures of the past, it is important to have a dedicated team of political leaders and bureaucrats under the Chief Minister to head the reform process, keep in readiness a dedicated early warning system and disaster management system that will quickly respond to the needs of the elderly, disabled and of the community in general.

(iv) Towards an Accident Reducing Road Policy

This priority may sound awkward in the present context. Even so, I have no difficulty in putting as a top priority a workable road accidents reduction policy as part of the Rebuilding project. We lost 433 precious lives in the floods of 2018. We cannot also forget that during 2018 nearly

43000 people were fatally injured and 4800 persons lost their lives. This is not something that we can happily write home about because we have been in this traffic situation for two decades now. We need to place a Road Reconstruction Policy that will reduce this everyday killing as part of our rebuilding project.

To conclude, it is to be understood that the quality of rebuilding of development depends on politics. Gramsci famously said that politics means conscious action (Praxis) in pursuit of a social goal. The rebuilding regime of political praxis in Kerala has to understand fully the meaning of this. A white paper and transparency guarantee document must be issued to win public confidence. Placating vocal voter interests can never be the way to building good and sustainable development. Reviving the spirit of renaissance means making public health and public education the centre piece of development as Scandinavian countries have successfully demonstrated. We have to make Kerala a thing of beauty and joy for the people of this planet. A dedicated team of chosen ministers and bureaucrats under the Chief Minster as an integral part of Social Development Council that makes the Nava Kerala Mission as its viable component can deliver the new era. There is the way, if there is the will.

Notes

1. It is important to note that in 2013 the FAO has designated Kuttanad agriculture as a globally important Agriculture Heritage System and is one of the 26 sites in the Asia Pacific Region.

References

1. **Sen Amartya (1999)**: *Development as Freedom,* Oxford University Press.
2. **Costanza, R *etal* (1997)**: 'Changes in Land Utilisation Pattern and the future of Agriculture' (in Malayalam), cited in George Thomas (2010) in *Bhoomi - a Public Property*, KSSP, Thrissur.

2

The 'New Kerala' Should be a Modern Welfare State

B. Vivekanandan

The Flood Disaster

Kerala is at such a juncture, engendered by a devastating flood disaster, and its terrible accompaniments, like a row of 537 landslides in the Western Ghats, caused by 12 days - long torrential rains all over the state, between 8th and 20th of August 2018. It has made all 44 rivers of the state in spate, and made all 42 major dams, and most of the 40 minor dams, in the state to lift their shutters to prevent more serious disasters. Reports say that 483 people have died and more than 57,000 homes and shops have been destroyed or damaged due to these floods and landslides. It has damaged about 10,000 kms of PWD roads, besides many bridges. In the agriculture sector alone, the loss suffered has been over Rs. 3646 crores. It is estimated that in order to restore the *status quo ante* the state would need more than Rs.25,700 crore.

To combat the flood havoc, the people of Kerala has shown a commendable sense of unity, cutting across all divides, which was manifest in abundance during the rescue and rehabilitation operations. A shining saga among them was the selfless devotion displayed by the fishermen of the Kerala coast, who have voluntarily jumped into the fray with their small fishing boats, their only means of livelihood. Their dare-devil actions during the rescue operations have already become a legend.

Partly Man-made

The flood disaster in Kerala, in August 2018, has been partly man-made. In 2011, an expert Committee on Environment, headed by Madhav

Gadgil, has specifically notified 17 areas of the Kerala region of Western Ghats, including Munnar, as ecologically fragile zones and recommended that this should be preserved undisturbed. But, in violation of this stipulation, vested interests were allowed to encroach and build up resorts and tourism centres, and other constructions, in these earmarked ecologically fragile areas, which has seriously disturbed the ecology of the Western Ghats. These activities were complemented by a large-scale stone quarrying all over the foothills of Western Ghats, in violation of these recommendations. The stone mining mafia gangs have grown up in large numbers and flourished everywhere, with the connivance of those who were in authority from time to time. The impact of mining activities in these quarries on the ecologically fragile areas of Western Ghats has been instant. When rocks in the foothills are mined by using powerful dynamites, it would inevitably shake the ecologically fragile areas at the top and also in the neighbourhood. Indeed, most of the 537 landslides in the Western Ghats, during those 12 traumatic days, took place due to this activity. Many people, and many homes, perished instantly due to these landslides. That is another facet of how this disaster has partly been man-made.

Other contributory factors which added to the intensity of floods have been filling up of fertile paddy fields in the state, to help the real estate businessmen to build high-rise buildings, causing shrinking of space for water percolation during the monsoons, and the blocking of natural outlets of water. The large-scale road-widening in the state from 8 metre wide two-lane roads to whopping 45-60 meters wide, 4-lane and 6-lane tarred roads all over the state, has also reduced the water percolation space. When natural outlets of water get blocked in such ways, by the overdrive of upstarts for luxury and pleasure, water would find its own way to flow, which can cumulatively take only the form of floods. These are other facets of how this disaster has partly been man-made.

In the circumstances, the government of Kerala should suitably rectify these mistakes, and get back to the Gadgil Committee recommendations and implement them faithfully and honestly, and stop all activities in the ecologically fragile zones of Western Ghats. It should strictly stop all stone-quarrying activities in the foothills of Western Ghats. Equally important is to accord overriding importance to environmental protection in all development policies of the government in future.

PWD Role in Restoration

In the restoration activities, a pivotal role should be assigned to public sector institutions. Transparency and public accountability in spending should be ensured to preserve public confidence in all pertinent spending. The repairing of damaged roads and bridges should necessarily be assigned only to the State Public Works Department, and, wherever necessary, also to the CPWD, to make the repair work corruption-free, and of high quality. The state PWD should be brought back to the centre of doing public works activities, and should be well-equipped, as a policy, to perform that role in future in a corruption-free manner. This is also an opportunity to strengthen the public sector in the state.

Fund Raising

In the fund raising drive to do the repair, the government should refrain from reckless borrowing from whoever offers loans to the state, without proper assessment of their intentions.

Need for Grants and Donations

The support Kerala needs today for rebuilding the State is not debilitating loans, but grants, cess and voluntary donations from inside Kerala, from other States of India and from abroad, particularly from reputed international solidarity foundations of Social Democratic Parties of Europe, which have a record of extending humanitarian help to victims of natural calamities. People of Kerala, and the Non-Resident Keralites abroad have responded positively to Chief Minister Pinarayi Vijayan's appeal for support to repair the damage, and to rehabilitate the victims of floods. In a spontaneous expression of solidarity with the people of Kerala, almost all State governments in India have responded spiritedly to the appeal of the Chief Minister. Many countries abroad have also indicated their intention to help Kerala to overcome the difficulties caused by the flood disaster. But, the Narendra Modi government, which is not familiar with, nor subscribes to, the principle of international solidarity, has thrown a spanner into such potential offers of open humanitarian support from abroad. But, the state leadership should strain every nerve to harness such support in forms of grants and donations for this purpose, from any part of the world. In a shrinking world, international solidarity actions are the would be norm in people-to-people relations in future.

Don't Take New Loans From World Bank

But, on the option of taking loans for the purpose, the government of Kerala should not rush for that easy option to overcome the present problems, and put the State's finance on the track of bankruptcy in the coming years. Since the floods in Kerala, we see the World Bank officials hovering around the government premises to thrust more loans on Kerala's shoulders by exploiting the present situation of the state. But, the government of Kerala should show the wisdom to reject such phoney overtures from the World Bank, and its associates, because its motives are malevolent.

Therefore, I strongly urge the government of Kerala to reject new loan offers from the World Bank. I would explain the context. Some years ago, I had the opportunity to meet Dr. Bruno Kreisky, former Chancellor of Austria, an outstanding world statesman and a staunch supporter of the development of the developing countries, in Vienna. During our conversation pertaining to the World Bank loans for third world development, Dr.Kreisky told me about the existence of a nexus between the World Bank and the giant automobile manufacturers of Europe and America, and about the motive behind the World Bank loans for the widening of motorable roads in the developing countries of Asia, Africa and Latin America. In an evaluation of future car markets in the world, the giant car makers of the West have found that, as car consumption in the developed countries of the West have reached a saturation point, their potential car market would be the populated developing countries of Asia, Africa and Latin America. They have assessed also that to promote a massive expansion of their car and lorry sales in future, a big widening of motorable roads in the developing countries would be a pre-requisite. The strategy they hatched, in collaboration with World Bank, was that, without spending any money from their pocket, they would achieve their objective of getting widened roads in all developing countries, for expanding their car market, through the provision of World Bank loans to the developing countries earmarked for the purpose. What does it mean? It means, the debt burden and the repayment liabilities are on the shoulders of the people of the developing countries, and the direct beneficiaries of it are the big car and lorry manufactures of Europe and the United States.

Kerala Government's ongoing road widening activities in the State, with World Bank loans, from its present 8m to 45-60 meters wide, under the garb of 'development' is basically to promote the car market interests of big car companies of the West, after making the people of Kerala and their children, to carry more and more debt burden on their shoulders. Should a Left government in Kerala also continue to support the market expansion interests of these big car companies of Europe and United States this way?

It has been reported that, during the recent floods in Kerala some of those newly widened roads in Kerala, built with World Bank loans, got damaged, and the World Bank officials have approached the Government of Kerala, with the advice to take more loans in the name of the people of Kerala, to repair the same roads, for the same purpose. Therefore, I urge the Left government of Kerala not to take more loans from the World Bank to serve the market interests of the car companies of the West.

Kerala's debt burden today is ` 1,86,453 crore. Adding to it with further loan from the World Bank, for expanding the car market of Western car companies, is, ideologically and rationally, not acceptable.

The 'New Kerala' Should be a Welfare State

Chief Minister Pinarayi Vijayan has declared that, by making use of the present opportunity, he would target the construction of a "New Kerala". But, he has not elaborated it. What he aims at by stating it, needs more clarity. Certainly, simply by repairing the damage that took place during the recent floods in the State, would not get established a "New Kerala". What we should strive for is to build Kerala as a modern Welfare State, which would ensure uniform social protection to one and all, from the "Cradle-to-the grave". It would mean that, a benign Kerala State would increasingly assume the main responsibility of the welfare of all people in the State through appropriate, public funded, institutionalised mechanisms.

It would also mean that the government would assume direct responsibility for the provision of social security, healthcare, medical care, education, social services, pensions, family allowances, child allowances, etc; for everyone in the State. And, all these provisions should be funded substantially from the state revenues. The system is in successful operation in all Scandinavian Social Democracies like Sweden, Denmark, Norway and Finland.

The Swedish Experience

In a country like Sweden, "the cradle-to-the-grave" Welfare State System works like this: The Welfare State benefits begin to reach a person as soon as a person is born. The sequence of it is like this. As soon as a child is born, the mother receives a gift packet from the government, with the official seal on it, containing baby clothes and other baby care items needed for the new born. After the child's birth, the parents begin to receive a child allowance, every month, for the healthy growth of the child. They receive this allowance till the child reaches 17 years old. The child's education is free; the study material is free; daily school lunch is free; and the public transport for all school going children is free. After the completion of education, it is the responsibility of the government to find him/her a gainful job, failing which, till such time a job is provided, the government is duty-bound to pay a maintenance allowance for him/her. When a person becomes a wage-earner, a reverse order begins and he/she becomes a tax-payer, thereby a financier of the Welfare State System. On retirement, a person gets about 65% of salary as pension, which means that the person can maintain almost the same standard of living after retirement. Healthcare of all Senior Citizens is free. Social Services provided to him are also free. When a person dies, the funeral/burial expenses are also met by the State. That is how the 'cradle-to-the-grave' welfare state system operates in Sweden.

Social security system in Sweden encompasses the right to work, right to education, a decent place to live, and so on. Under the system, the entire population is insured against illness. Everyone gets equal opportunity for education. The system protects all employees from work related accidents and sickness. All employees are entitled to compensation for industrial injury. Housewives are entitled to annual holidaying at any tourist spot in the country at the governmental expenditure. All mothers receive maternity grants. Under a family law, either father or mother is entitled to 7-monhs leave, with full pay, after the birth of a child.

There is highly developed pension scheme in Sweden. Under the old-age pension scheme, all old people are entitled to a pension. Similarly, all disabled people, after the age of 16, are entitled to a pension. If a person is fully disabled, he/she would get a full disability pension, which is equal to the old-age pension. Similarly, parents are entitled to childcare allowance, if they themselves nurse their disabled children.

In Sweden, medical care is free, except a nominal prescription fee.

A notable feature of the Swedish welfare state system is that it is attuned to promote egalitarianism in society. The application of the Steeply Progressive Income-tax system, and the solidarity wage-structuring, has substantially reduced the income disparity between the take-home pays of the low-paid and high-paid employees.

Indeed, it is worth sending a few experts from Kerala State Planning Board to Sweden, to study and report the nitty-gritty of how the Welfare State System is organized, operated and funded there, as an input to make Kerala as a Modern Welfare State.

However, I hope, the Kerala Chief Minister would attune his "New Kerala" concept to make Kerala a Welfare State, based on equality, equal justice and solidarity, in the coming years.

3

Citizenship and Reimagination of Keralam

P. Sanal Mohan

Citizenship is the most decisive factor that determines the quality of life in modern democracies. The institutional paraphernalia of the governments help deepen the idea of citizenship through their practical policies aimed at transforming the lives of the citizens. The idea of citizenship becomes much more complex under such situations than the conception of citizenship as an individual's belongingness to a polity. Although the idea of democratic citizenship is egalitarian, the structured and hierarchized practices have defined the everyday life of citizens in many modern democracies.

In India the idea of citizenship gained ground through the anti colonial political mobilization of the twentieth century. The last seven decades of postcolonial democratic experience has placed citizenship as a serious political issue. Often thought of as a settled political question, several aspects of citizenship are contested in India today, especially when delivery of social goods and welfare activities targeted at different social groups is considered. The citizens confront a variety of problems in their everyday life in accessing the social goods granted by the state. One could even say that to large segments of the population the state is experienced as an everyday phenomenon influencing their lives. Therefore the problems of citizenship have become serious problems to be confronted although often times they are not recognized in their complexity. From the late twentieth century onwards the idea of citizenship was stretched to its limits in India thanks to different social movements especially that of Dalits and Adivasis. It is appropriate here to mention the environmental movements in India that challenged the official wisdom and offered a critique of the reigning ideas of economic development that was at the cost of forests and other natural resources that can't be reproduced or that are non-renewable. Such movements, protracted in character, may assume significant momentum in the wake of challenges unleashed by powerful events such as the climate change induced 'natural disasters'. Keralam is not an exception to this general situation of the social movements although the dominant social classes in Keralam often deny such a phenomenon. It is possible that in the twenty first century the environmental problems might give rise to social movements that could bring together disparate social groups and classes. However, it is generally acknowledged that the poor and vulnerable sections become the worst victims of natural calamities and effects of climate change as they do not have necessary resources to fall back up on.

This paper seeks to analyze the questions of citizenship particularly probing the experiences of the marginalized sections of Keralam as it became decisive in the context of 2018 floods. Their efforts to survive the flood and its aftermath were severely constrained by the lack of adequate resources which were related to their conditions of existence in Keralam itself a product of their own history. One of the important historical reasons for this has been their alienation from property. In liberal democratic theory, ownership of property is fundamental to the idea of citizenship or citizenship comes along with property. However, in India citizenship and exercise of the rights of the citizens were tagged to ownership of property only in the colonial times. While it is important to divest property qualification from citizenship, ownership of property can't be undervalued in India where historically large segments of the people were prevented from owning property due to the religious ideology of caste. In Keralam the institution of caste slavery prevented the slave castes from owning land. A considerable number of Dalits, both Christian and Hindu with slave caste ancestry were structurally prevented from owning land. These were the castes that provided physical labour in Keralam historically. If may be reiterated that their social and economic vulnerability was directly linked to the historical antecedents that have not been adequately analyzed and understood in Keralam. This has had disastrous consequences for these social groups, large majority of whom were agricultural labourers who experienced a raw deal at the hands of the government of Keralam led by the communists when land reforms were implemented. Large majority of the agricultural labourers from the Dalit communities received only homesteads, as they were not rent paying class of tenants. This homestead of lands was not enough for them to evolve as independent peasants. The extent of these homesteads was never more than ten cents in villages and five cents in towns and municipalities, which became overcrowded as decades passed by and population grew.

It may also be stated that peasant classes that are visible among Dalits were largely due to their migration to the hills of the former native state of Travancore or to the hills of Malabar. Even in these instances we may not come across a class of middle peasants emerging. Instead we may count them only as small and marginal peasants as the ownership of land was on an average less than three acres.. However, the important point to be noted is that such efforts of Dalits owed in most cases to their own initiative and not due to the support of the post independent state.Those Dalits who moved out of their ancestral villages mostly in the native state of Travancore,could develop as peasants and were able to claim their rights as citizens in the process against several odds in the late 1940's and early 1950's. It may be noted here that in some cases the Dalit Christians could develop their social movements as they could own landed property and become small and marginal peasants through migration and thereby liberating themselves from the dependence on the upper castes, both Christians and Hindus. In this example the crucial point is the ownership of property that allowed the hitherto enslaved Dalit Christians to articulate their agency. In other words, in caste society the idea of citizenship will remain a forbidden fruit unless it is welded to the ownership of property for the oppressed castes. It is in this context that I wish to consider citizenship and property in the context of caste. It is necessary in this context to point out that when discussions on the economic and social development of the slave castes were discussed after the abolition of slavery even in the 1870's, the officials of the Travancore and Cochin native states felt that ownership of landed property was essential to develop them as 'civilized' people. I shall mention further that right from the mid nineteenth century onwards when the Protestant missionaries began to work among Dalits they had observed the need for landed property to liberate the slave castes of Travancore and Cochin. It was equally true for the slave castes in the British Malabar too. We may cite here an example from the history of a late colonial Christian missionary organization. From the late 1930's when the Jesuit missionaries laboured among the Pulayas in Chirakkal, a few miles north of Kannoor town, one of them observed that..." if we want to make out of these wretches, good Christians and decent and respectable citizens... we must aim at, try in course of time to enable them to live in a decent house of their own; to possess a little piece of land something anyhow, that may be the foundation, the starting point, for them to rise rung by rung in the ladder of social life"[1]. In the second decade of the twenty first century now in the background of floods and their aftermath the question of ownership of property becomes crucial.

Keralam's tryst with floods in July-August 2018 posed unprecedented questions on several aspects of the social and political life of Keralam. These questions are important in the larger context of the efforts of the Government of Keralam to initiate reconstruction of Keralam that is referred to as 'Construction of New Kerala'. Spurring new thoughts on social life in Keralam, the phase of reconstruction as announced by the Government gave lot of reasons for the affected people who had been historically marginalized to expect something very new. In other words, the post flood situation demands reimagination of a new Keralam. Such a situation demands a critique of Keralam's social and economic development, famously referred to as 'Kerala Model of Development' as it evolved historically. The challenges faced in the post flood situation unsettle the prevailing consensus on citizenship and its various manifestations in Keralam.

Let me analyze the post flood situation by studying the experiences of the marginalized sections. In the subsequent sections of the paper, I argue that the deficit in citizenship that the marginalized sections had experienced historically becomes complicated when faced with the fury of natural disasters. The marginalized sections do not have access to resources to overcome destructions caused by disasters of great magnitude such as that happened in July-August 2018. In order to explore possibilities of resolving such situations similar to that happened in the 2018 July floods, we need to begin with a radically different notion of citizenship.

The Need for Reimagining Keralam

Imagination is a word often used in critical social science to buttress the argument that larger collectivities such as nation states or for that matter any community, gets formed in the process of them being imagined by people, which is a long drawn out process. Today this would appear as a statement of the obvious. However, we live in a particular context in which even such reiterations become politically and academically significant as people attribute immutability and eternity to prevailing institutional forms and social practices. More importantly, it needs to be reiterated that such forms are liable to fundamental changes and also that they were constituted at particular historical moments. It is in this context that I argue that the value-loaded images of Keralam that are in circulation today need critical analysis to find out the momentous flaws that were often glossed over in the euphoria created by such images. The political community in Keralam, has reached a situation where in, it is no longer possible to move ahead with the ideas of Keralam that were uncritically accepted although some sections of scholars have been critical of such a perspective. Most notably, the critiques were directed against the idea of "Kerala Model of Development". While it is important to preserve the gains of postcolonial transformations of the society, polity and economy, with their roots in the colonial transformations, it is necessary to think afresh critically about the challenges ahead. It is in this context that the question of reimagination of Keralam becomes imperative. Now let me open up the question of reimagination. I have used the word reimagination in this paper as much of the discussion in Keralam in the post flood situation hovers around the problem of construction of a new Keralam. It is this idea of building a new Keralam that makes it necessary for us to think about reimagination of Keralam. Much of it has to be done by relocating ourselves in the margins of Kerala society. Following Homi Bhabah, I will argue that margins are places where things begin and not spaces where things end. Therefore, any idea of new Keralam has to begin from the visibility afforded by the margins. In fact this is nothing new as scholars have been focusing attention on the limits of the hither to achieved social and economic developments of Keralam. The critiques of Kerala model of development have shown its limits reiterating the fact that the fruits of developments in Keralam did not benefit the adivasis, dalits and the fisher people. Likewise it is argued that the developments did not benefit women of certain social groups. If the pattern of social and economic development followed in the state did not benefit significant sections in Keralam, it would mean that there were a number of crucial factors that prevented the marginalized from accessing the fruits of development or the benefits of development reaching them. However, these questions were never problematized. Therefore, in the current situation when we reimagine a new Keralam we have to address the problems of the marginalized communities and social groups in Keralam. It is in this context that I wish to problematize the idea of citizenship in contemporary Keralam.

While there is an agreement about the fact that the idea of equal citizenship was very central to the development of the historically oppressed social groups in India, it has been observed by scholars that the attribute of citizenship was not given out to all social groups equally.

In spite of these differences, the idea of citizenship is fundamental to the further development of the oppressed social groups. This is spectacular in a situation in post colonial India where civil society has become 'closed and completely elitist.'[2] Closely following this is the crucial distinction made of real and formal citizenship. Here in fact equal citizenship becomes the real issue. This according to Chatterjee is due to a fundamental shift in the manner in which democracy was defined historically. He proposed the path breaking concept of the politics of the governed that has analytical potential in contrast to the traditional definition of democracy. A significant aspect of the postcolonial history of Dalits and other historically, oppressed groups has been interpreted by historian Gyanendra Pandey as their conversion to citizenship. However, he also notes the limitations of this idea of citizenship[3]. Developed in western political theory and practice, the ideas of democracy and citizenship have been naturalized in the colonial world following the long nineteenth century. India is a classic example of this intricate process. Without going into the complex history of it, I wish to state that not everyone was equally poised to evolve as citizens. It is here that we come across the problem of communities that remained as privileged ones and communities that continued to exist with their historic "backwardness'. Both these began to pose serious questions to our understanding of citizenship vis-a- vis modern state. Modern state is the site where these contradictions are to be resolved. However such conflicts were never resolved fully, and the civil society became the arena where such conflicts were articulated. Needless to say that civil society became increasingly elitist and exclusive. It may be argued that a similar process was at work in Keralam also, that made the civil society in Keralam in the postcolonial times particularly elitist.The oppressed caste movements that contributed substantially to the development of modern civil society in Keralam, through an inversion of history, got alienated from civil society and became, in turn, almost external to it. In the absence of social movements articulating their problems historically in the civil society, it was difficult for the Adivasis to evolve as modern citizens. Probably it was in the last decades of the twentieth century and in the first decade of the present century that the citizenship of Adivasis became a central issue in the politics of Keralam. The point to be noted is that the substantive issues affecting the community had to wait for such a long time to be articulated. This showed the deficit in the democratic citizenship of Adivasis who were historically alienated from the agrarian society but integral to modern system of power represented in the plantation economy in Keralam. Their alienation from the agrarian society does not mean that their labour was not used in agricultural production in many parts of Keralam. Nevertheless, there is a curious way in which their ethnic identity got merged with the plantation working class identity that posed serious questions to their social existence as modern citizens. In a way we need to ask the question; what are the ways in which the stigmatized social groups could claim citizenship in modern society? We may identify a similar situation in the case of the fisher people in Keralam.

Historically at the margins of the land system and opened to the sea, the fisher people had a liminal space open to them. On the Keralam's coast they had a history of being open to external influences though they might have existed in a reclusive manner when it comes to their relations with the agrarian communities. However, in the twentieth century we observe them developing social movements that critiqued the prevailing notions of domination based on caste. I wish to argue that the formal aspects of citizenship, which was extremely technical, prevented a substantial section of the oppressed people from realizing substantive social citizenship. This created a situation of their becoming increasingly alienated.

In the postcolonial social and economic development, Dalits and Adivasis got limited to their particular enclaves that demand an exploration of spatial aspects of citizenship. It may be argued that modern democratic citizenship of postcolonial times could not resolve the historically accumulated problems that the marginalized sections of Keralam experienced. Therefore inspite of much hyped transformation of Keralam they remained vulnerable. The problems of marginality in terms of public sphere and spatiality were never addressed in Keralam.

The inadequate conception of citizenship has made matters worse for the marginalized sections that could not develop themselves as a substantial economic class. The same reason might perhaps explain why a powerful middle class did not emerge from their rank and file. It is here that issue of social citizenship become quite crucial. What are the implications of social citizenship in Keralam? Ideas of social citizenship would require the creation of more egalitarian structures that will address the problems of the disadvantaged groups although liberal ideas of citizenship may critique this as it provided more room for the intervention of the state through welfare policies. In spite of this it is important that state support is extended to under privileged sections which is a concern of social citizenship.

Spatiality of Social Existence

I wish to further argue that the spatiality of physical and social life as it evolved in Keralam was crucial in the post flood rehabilitation and relief works. This situation was a product certain history of spatiality in Keralam. For example most Dalit settlements in the wetland regions of Keralam are located on the narrow bunds of the paddy fields. In many such places in the Kottayam and Alappuzha districts several hundreds of houses were completely submerged under water and the people had to be rescued and sheltered in camps opened in schools and other public buildings. We may recall here the issues discussed in the previous sections of this paper that historically those from the slave castes who were harnessed as suppliers of physical labour in the wetland system were settled on the embankment of the paddy fields and that form of settlement continues even today as land has become costly and scarce making it difficult for people to move out even if they want to do that. One might also mention that their long historical connection with this particular geographical setting would prevent them from moving out. Yet, we also have examples of thousands of people from the slave castes who were part of wetland rice production moving to the hills and settling there as agriculturists.

In places where the government offered houses to Dalits and other marginalized sections through the programme of One Lakh Houses, such settlements were situated mostly on lands that were not really suited for agriculture or lands that were of inferior quality particularly in terms of availability of water and other basic facilities. Amounting to spatial marginalization, this had deleterious effects on the people thus relocated. Another crucial aspect in relation to social life in Keralam is the spatial identification of individuals and families as expressed in their family names/house names. Large majority of the oppressed slave castes derived their house names from that of their masters. Often times family names have been derived from features of land or trees or other natural elements similar to a totem, to use an anthropological expression, which is typical of agrarian social life. These house names become markers of cultural significance that distinguishes individuals and families locating them in the specific socio-cultural space. This particular aspect of family names/ house names could constitute an important aspect of cultural citizenship. It is again important to note that even people without much landed property can also have a house name connoting to land. These examples have been mentioned to show the significance of land ownership and property in the evolution of citizenship in Keralam, which is distinct in certain respects from property qualification that decided franchise historically.

When it comes to post flood reconstruction it is necessary to consider these aspects of social and cultural citizenship that is quite fundamental to Keralam. The question of spatiality is crucially connected with the fundamental question of human dignity. It is necessary to ask if in the situation of floods and their aftermath we encounter issues of human dignity articulated specifically reminiscent of the phase of anti- caste mobilization in twentieth century Keralam.

Human Dignity

Dignity of the oppressed castes was one of the central questions of the anti caste social movement of the twentieth century. We find the same ideas being expressed in a forceful manner in the writing of missionaries who worked among the Pulayas and other slave castes. According to a perceptive missionary, "the most important thing was to treat the oppressed caste people as human beings, respect them and their rights and also to give them their just wages"[4]. It may be noted that the missionary had brought together the problems of the symbolic and the economic aspects of social life together in the analysis of caste question. Further, we should also understand the continued significance of human dignity as a contested problem even in the twenty first century. I may mention here the protracted land struggles in Keralam beginning in the last decade of the twentieth century spearheaded by Dalit and Adivasi organizations in which the question of human dignity is second to none. I wish to elaborate a few problems related to the problem of space in the context of our concern with human dignity. Historically, the slave castes in Keralam lived in small huts, which one 19th century observer compared to a large basket turned upside down. Ethnographic information collected in Dalit settlements also testifies to the fact that even in the mid twentieth century the large majority of them lived in small huts built of thatched roof which was supported by four poles. There was hardly any idea of separate rooms and other amenities, which would mean that there was no question of built up space available to people. This would also mean that the idea of privacy may not have developed in a big way among them. A dignified life would require access to built up space that is necessary for life to flourish.

If we analyze the manner in which space was controlled in the caste system - separate and distinct settlements of slave castes away from that of the upper castes without access to water, roads and other facilities - soon we realize the ubiquitous nature of spatial inequality. In a way, dignity of individuals and social groups was denied by monopolizing the space. The same practice could be reproduced when it comes to bodily practices, control over spaces of social interaction and the higher levels of aesthetic creations and knowledge production. It may be noted that the separate settlements that are euphemistically referred to as colonies in the post independence phase where the occupants are mostly from the oppressed castes give an extreme example of spatial inequality, which undermine human dignity. Over the decades there developed a sort of ghettos out of these settlements. It has been reported that in many places such settlements were submerged during the floods of 2018 and people had to be relocated in the government run relief camps. These camps became the sites where spatial inequality was reproduced making it impossible for people to associate with one another freely as they were conditioned by certain fixed notions of exclusive social space. Is it possible to think of relief camps as liminal spaces where people could overcome pre-existing notions of spatial inequality? However, in the narratives that were widely circulated we understand that although people were ready to sink their differences in the face of calamities brought in by the floods, the differences soon returned to haunt them once they were forced to share space.

In the 2018 July - August floods thc pre-existing forms of spatial divisions became more explicit although spatial inequality has always been there in Kerala historically. This particular idea of spatiality got worked out in fields as diverse as ownership of land, agriculture, use of natural resources, construction of houses, sartorial culture, food, sexuality and affective dimensions of human life. It may be mentioned that several aspects of these spatial configuration went into the making of what is often referred to as Malayali cultural practices. Hegemonic in nature, these cultural practices that defined the general features of social life in Keralam was challenged by anticaste movements beginning in the twentieth century although they had some precedent in the mobilization of the protestant missionaries against slavery and spatial inequality. However, several aspects of the dominant culture got normalized and eventually integrated with the value system of the emergent middle class in Keralam which continue to maintain caste and class privilege. Most recent challenges to this consensus could be seen in instances of land struggles in various parts of Keralam as mentioned before. One possible question that could be asked is the relevance of this particular argument to the proposed reconstruction of Keralam. While the construction of new Keralam as a significant political project, unless we offer a critique of caste/class privileges in Keralam that made ubiquitous forms of domination as normal, it would be well-nigh impossible to create a new Keralam.

New Keralam has to be thought of in terms of some radical alternatives along with reconstruction of houses and new built spaces. Needless to say that reconstruction of damaged and destroyed houses is important and essential for the well being of the people who have become homeless. At the same time what is more important is the possibility of imagining Keralam as a place where more egalitarian modes of living could flourish. Such a mode of living could be imagined by bringing into consideration the question of recognition. In modern social theory when the question of dignity is analyzed particularly in the context of structural inequalities, often scholars emphasize the need for recognition. People who are oppressed on account of various reasons - historical and contemporary - are to be recognized in terms of their essential humanity which alone will bestow dignity on them. What is the significance of this in a caste society? In a caste society there are structural instruments that limit the reach of the idea of dignity. In other words deficit of dignity attributed to the oppressed caste people can erode their humanness, which in turn can create misrecognition. This question will take us to some of the fundamental issues related to the current phase of social and economic changes in Keralam as well as the historical evolution of different social classes in particular a middle class in Keralam. There prevails in Keralam a situation of misrecognition of the oppressed caste people that could fatally undermine their sense of dignity. However, if we follow the narratives of the oppressed castes shared within such communities we get a different idea of human dignity. It seems following ethnographic evidence that in spite of misrecognition slave castes could retain their sense of dignity which was articulated through a range of religious and social mobilizations of the late nineteenth and twentieth centuries.

Closely following this, I wish to identify the ideas of social justice. Usually social justice is a notion that tends to offer a corrective approach to certain historical wrongs. Yet, the idea is not to repeat the historical mistakes or oppressions but to guard against all forms of domination and violence in future. Therefore, today social justice can't exist as a stand alone concept, but it has to be thought through along with issues of environmental justice very seriously. There were times when people used to debate social justice and environmental problems as trade offs.

However, the 2018 floods and their aftermath loudly proclaimed the need for considering environment problems as much as we consider the needs of the marginalized sections of society or broadly the needs of human society. Such a realization dawns on people today as we observe the worst sufferers of natural calamities including floods are the marginalized groups who are without much resources to cope up with such challenges. Their vulnerability is bound to increase in future as we face the staggering truth of climate change. In addition to this, when we consider other aspects of floods it is true that it's effect was not confined to certain sections of the society alone but in areas where it turned out to be a deluge, it affected everyone. But the historically marginalized sections did not have the required resource base to overcome the post flood situation. It is in this context that we need to address historical dimensions of spatial inequality that are to be seriously considered.

Conclusion

The massive efforts of the government and society in Keralam to reconstruct the state in the aftermath of the devastating floods of 2018 July - August is highly laudable in spite of the financial crunch and several other odds facing this Herculean task. Yet, we find the strides towards a better future for the flood-affected people in spite of several shortcomings noted in the implementation of the proposed schemes. It is here that we in Keralam are to be guided by a different notion of citizenship, which will enable the government, and people at the helm of affairs to approach the flood affected people from a different perspective. They are not target groups of policies of governmentality, but they are to be treated as citizens claiming to substantive aspects of citizenship such as cultural and social citizenship discussed in this paper. Capable of providing a different idea of rehabilitation, such a perspective will help overcome issues of spatial injustice and inequality that has a long history in Keralam which was aggravated in the context of the flood and its aftermath.

Similarly the preparedness of the state to meet eventualities should be guided by the ideas of recognition and dignity, which are inalienable aspects of the lives of the people, whether rich or poor. Deficit in the idea of dignity was visible in the post flood situation when the relief camps were closed and in many cases people were asked to return to houses, which were unsafe. On many occasions the state agencies were unwilling to accept the judgments of the affected people and the estimates of the damage by the former fell short of the actual losses people suffered.

This happened mostly in the case of people from the marginalized communities. There are several significant practical problems that people who were relocated to relief camps had to face and one of which was shortage of toilets. People and governments in Keralam boast of the high quality of life in Keralam but when it comes to sanitation the claims of the people as well as government appear to be hollow.

One central question here is the inability of the people to transcend certain mindsets that have been the creation of caste world view that does not find anything wrong in the manner in which filth is treated in Keralam. Along with this we could also talk about the negligence of public rest rooms and other facilities in Keralam. This actually points to a deficit in the understanding of people in Keralam regarding the dignified life of citizens to which all are entitled. It demands an aesthetic approach towards such issues of everyday existence, which alone can resolve the outstanding issues of waste management in Keralam. One major problem discussed in the camps has been inadequate number of rest rooms along with lack of running water and proper cleaning of the facilities. In fact it is a major point that shows the failure of civilized existence in Keralam. Even in normal times there will not be sufficient number of clean rest rooms in public buildings and public spaces that are converted into relief camps during natural calamities such as floods. These are problems that have never been addressed properly in spite of the tall claims about social and economic development in Keralam. I have taken a single example to show how the state fails in one of the most important aspects of dignified life of people in an everyday context. Let us add on to this medical emergencies and a host of other questions where dignity of the poor and marginalized sections are violated on an everyday basis even in normal conditions. Therefore, any thinking on New Keralam will naturally have to address issues of citizenship, recognition and dignity of the people particularly that of the marginalized sections as the core issues that will transform social life in Keralam.

Notes

1. Rev. Fr. J. Taffarel SJ (1950): *Jottings of a Poor Missionary*, The City Press, Mangalore, pp.103-4. Emphasis in the original.
2. Partha Chatterjee (2004): *The Politics of the Governed: Reflections on Popular Politics in Most of the World*, Permanent Black, New Delhi, pp.4-9.
3. Gyanendra Pandey, "Time of the Dalit Conversion" *Economic and Political Weekly* Vol. 41, No.18 (May 6-12, 2006), pp. 1779+1781-1788
4. Rev. Fr. J. Taffarel SJ (1950): *Jottings of a Poor Missionary*, p.109

4

Rebuilding Kerala: Lessons from Aceh

John Kurien

The deluge in Kerala and the challenge to rebuild a 'New Kerala' reminded me of my four-year involvement in post-tsunami efforts at Aceh Province of Indonesia to 'build back better' - the phrase coined by US President Clinton when visiting Aceh.

I learnt in Aceh that disaster management has five phases: rescue, relief, rehabilitation, reconstruction and redevelopment. The first four are part of common discourse in Kerala. By redevelopment, however, we are alluding to using the context of a disaster to re-envision a change in the composition, pace and direction of the development process.

Aceh Province bears close resemblance to Kerala in physical features - west facing coastline; hills in the east running parallel to the coast; and a plenitude of rivers flowing westwards. Some common socio-cultural traits make a Keralite nostalgic. They include the '*kadai kopi*' (coffee shop) with the single newspaper and contentious political arguments in the morning. Many Acehnese trace their origins to Ponnani in Malabar.

In Aceh, the monster 8 metre tsunami wave of December 2004 swallowed over 150,000 persons in less than 30 minutes, often reaching over 5 kilometres inside the coastline. It totally wiped out coastal towns and villages, leaving the traumatised survivors in a state of shock.

In Kerala, the well-coordinated and heart-rending rescue phase is over. Relief emanated from every corner of the state, country and abroad. Undoubtedly, the best in us manifests in the worst of times. After the flood waters receded people went back to their silt and mud filled homes. Rehabilitating people back to their homes is marked by an upswell of universal goodwill and fraternal feeling without any consideration of the barriers and identities that divide us in good times.

The narrative of the discourse at all levels - affected households, the local communities, and the various tiers of governance - has shifted to thinking about the priorities and strategies for reconstruction. How can Kerala turn this crisis into an opportunity of sorts?

Leadership for Reconstruction

In Aceh, reconstruction was coordinated by a specially created agency, decreed by the President of the Republic of Indonesia. The Rehabilitation and Reconstruction Agency - BRR for short - was designated to operate for a four-year period, with competent full-time staff and two oversight boards. The Agency's mission was to "restore livelihoods and strengthen affected communities by designing and overseeing a coordinated, community-driven reconstruction programme implemented according to the highest professional standards."

There was a Master Plan for Rehabilitation and Reconstruction which matched donor funds to specific community needs through a process that was rigorous, sensitive to local concerns and priorities, and well-monitored. BRR coordination was essential to ensure timely delivery of support and eliminate legal and institutional uncertainties in reconstruction.

BRR recognized the important role of local community input and participation in reconstruction. Local government bodies, customary and community leaders, academics, religious leaders and NGOs were continuously consulted and involved in re-shaping the reconstruction plan whenever needed. Mechanisms to trigger community-based reconstruction initiatives were critical to ensure participation - and more importantly to eliminate perception of bias or corruption.

The lesson for Kerala is that reconstruction is best achieved with a separately designated, time-bound and focussed entity for execution, with sanction of the legislature and accountability to it, but with the freedom to function according to the highest professional and ethical standards. This entity should be viewed as a public-private-people partnership for building back better.

Its mission should be to create a part fixed, part flexible physical master plan from a judicious amalgamation of participatory ward/panchayat level physical rehabilitation and reconstruction plans. Each ward/panchayat should provide a vision statement for their rebuilt reality. This amalgam must then be meshed with macro-assessments of reconstruction needs which have taken seriously the geo-physical and environmental factors in nature made amply visible in the wake of the deluge.

Aid Coordination

With aid pouring into Aceh (it reached USD 8 billion), the Government of Indonesia requested the World Bank to establish a Multi-Donor Trust Fund (MDF) to pool donor contributions. The MDF activities had to be consistent with, and guided by, the Government's Master Plan and under the leadership and direction of BRR.

The MDF helped all donors allocate funds judiciously. It provided detailed spatial and financial information on the specific needs of the survivors. Procurement procedures were made simpler and transparent shortening implementation time. Smaller donors with limited overseas experience or administrative capacity also used MDF as a channel for their aid.

In Kerala, a multi-donor profile is gradually evolving for assisting reconstruction. Aid from central government, foreign nation states and international development agencies and banks have fixed arrangements and channels to reach the coffers of the state. Kerala can also make project proposals for specific short and long-term needs, availing of soft loans and grants from a range of sources including the World Bank and the United Nations system. The Kerala diaspora is ready to answer the Chief Minister's call to pledge a month's earnings for this cause. They will pledge more if the state can guarantee that hard-earned money they contribute can, if they so indicate, be utilised for specific projects intended for designated places and people.

Kerala needs a special purpose financial trust facility for overall and nuanced aid coordination. This facility should have the Chief Minister and Finance Minister at its helm, but the operational charge needs to be given to qualified bureaucrats and experts with experience in dealing with the financial management of disaster reconstruction.

Linking Aid to Execution

In Aceh, when dealing with the post-tsunami relief, information technology and local expertise was not as developed as in Kerala today. The BRR depended on e-mail, SMS, satellite phones, GIS, and regular ground-checks. Matching funds and personnel to exact locations for reconstruction work took time due to complete destruction of road infrastructure. Delays gave rise to doubts about credibility of the process. Local level coordination and execution were dependent on physical meetings with paper charts. At BRR there was the occasional power-point presentations and rare video conferencing arrangements.

Despite these shortcomings, credibly linking aid and execution helped form strong partnerships and investments were thereby effectively utilised. Information about quantum, location and utilisation of aid was made more transparent.

In Kerala today, the ubiquitous smart phone, provides transparency, accountability and empowerment. Citizens photos of damage or progress of restoration can spread through social media networks. Our talented IT personnel, equipped with open source and open hearts, can make a one-to-one match between anyone's donation with the specific reconstruction needs of a village, a people or special cause of her choice. The twinning offers both donor and recipient, transparency and accountability at almost zero marginal costs. This enhances the trust of individual and institutional donors. The result is more aid.

Tsunami of Private Aid

The Aceh tsunami was the first major disaster seen worldwide on live TV reportage. The impact on hearts and purse strings of viewers was without parallel. In Aceh, after the tsunami their next problem was the tsunami of aid.

Direct private aid, channelled through local and religious institutions were hard to stop, but they created their fair share of contentious issues on the ground. Allegations of corruption, unwarranted compensation to those who suffered less or did not suffer at all, were rampant.

In Kerala too, NGOs and religious organisations have their clientele and obligations towards them. They may not be willing to pool their resources into any common kitty, particularly if controlled by the state. It is hard to trample on the rights of such organisations. However, these acts of kindness and favouritism, must be recorded at the local panchayat level, to discourage their beneficiaries making duplicate claims to aid from public funds. Affected households, from upper echelons of our society, may not lay claim to government financial assistance. They need to be separately registered. Along with others they will require assistance for getting duplicates of legal documents, linking back to the electricity grid and road network. Thus, dealing with the official reconstruction apparatus becomes inevitable.

Redevelopment Challenges

For Aceh, the 'gift of the tsunami' was total provincial autonomy granted by Government of Indonesia in matters of political governance, customary, cultural and social expression and management of natural resources, among others.

The Government of Aceh became open to a new development paradigm for natural resources giving serious consideration to co-management - where state, community and market - coalesce from the lowest level upwards to protect and utilise the real wealth of society for a just, participatory, sustainable and self-reliant process of development.

In Kerala, our settlement pattern, food grains, cash crops, tourism, power sources, fisheries, and easy access to fresh water are a few essential attributes which arise innately from the special geo-physical contours of the state. The recent floods have highlighted once again the inevitable need for greater partnership of state and community to utilise and manage these precious natural resources. We must utilise the golden opportunity, thrown up by crisis, to re-imagine and re-design how we situate ourselves within the context of Kerala's salubrious natural environment.

In the context of global climate change, if we do not wish another deluge, let us forge a measured balance between 'deep ecological activism' for conserving nature on the one hand, and on the other, a well-planned reconstruction of our habitations, our choice of cash crop agriculture, a re-look on our dams, a willingness not to disturb the banks and sand of our rivers, and an agreement to leave our beaches as playgrounds for the sea.

To achieve this, we need, like Aceh, a renewed political engagement, in a new framework, where our commitment is not to narrow political, class and caste identities but open to envisioning and co-creating a new Kerala, within the purview of its natural boundaries, with the pledge to build back better.

5

Social Protection in Post-disaster Situation in Kerala

Udaya S. Mishra

Introduction

Social protection is one of the foremost concerns of the state government to ensure that vulnerable population gets appropriate and effective support to overcome financial insecurity. All countries have their own social protection schemes. In India, most of the social protection schemes are implemented through the state governments or local self governments. In fact, the efficiency of the state governments along with the institutional infrastructure towards its execution really determines the success of the programme implementation. In this paper, an attempt is made to reflect on the significance of the social protection mechanism in Kerala in the context of recently happened disasters.

Social protection generally constitutes policies and programmes that seek to reduce poverty and vulnerability of those at risk and enhance the social status and rights of the marginalized by promoting and protecting livelihood and employment, protecting against hazards and sudden loss of income, and improving people's resilience to withstand risks.

The major components of social protection are labour market programmes, social welfare programmes, social insurance, and social safety nets. The labour market programmes constitute the measures that are aimed at enhancing employment opportunities and protection of the rights and welfare of workers. Preventive welfare programmes are basically the preventive and development interventions that seek to support minimum basic requirements of the poor. Social insurance seeks to mitigate income risks by pooling resources and spreading risks across time and classes. Social safety nets are stop-gap mechanisms or urgent responses that address effects of shocks on specific vulnerable groups.

The need to strengthen social protection has been renewed in the post-disaster period. In Kerala, the disasters have made severe damage to the individual well being, and it may have serious implications towards impoverishment of a large section of population. This situation should be prevented at any cost. Because, once the population confront poverty, it is rather difficult to re-build their socio-economic profile equivalent to the pre-disaster situation. Again, the poverty consequence is muti-dimensional in nature. The risk arises from the disaster and is all pervasive. It touches all spheres of livelihood assets particularly the physical capital, natural capital and financial capital. Unless government take proactive actions by strengthening social protection measures, it would be very difficult to mitigate the distress condition aroused from these disasters.

In this context, it is necessary to revisit the implementation of social protection programmes in Kerala, and rework it if required and make sure that social protection alleviates poverty among the vulnerable sections in the state.

Risk and Vulnerability

During disaster situations, vulnerabilities can emerge at multiple levels as shown in Table 5.1. From past experiences in different parts of the world, it is realized that risks usually associated with disaster may be broadly classified as individual risks, economic risks, environmental and natural risks, and political and governance risks. Each of these has its own vulnerabilities as mentioned in the Table 5.1 and the coping strategies required by the state differ accordingly. If the disaster is new to the population, the risk would be severe for the people and takes time to overcome it. In the case of Kerala, the recent floods was an entirely new experience for the present generation and this unexpected event has made severe damage to family's livelihood giving rise to vulnerability that can have serious potential for many people landing in miseries of poverty.

Table 5.1 : The social risks and associated vulnerabilities at the time of post disaster situation.

Social risks	Vulnerabilities
Individual life cycle	Hunger, illness, disability, old age, death
Economic	Unemployment, underemployment, low and irregular incomes, economic crises
Environmental and natural	Typhoons, drought, floods, earthquakes, volcanic eruption
Political and governance	Armed conflicts, crime, corruption, social exclusion and discrimination

In post disaster situation, it has also been noticed that the magnitude of risks varies across different sections of population, and the risk is more among the vulnerable sections like children, and old aged population in general. However, women and marginalized groups are the most affected by the risk factors. Table 5.2 shows the major possible social risks that can be confronted by various vulnerable groups in post disaster situation in any population.

Table 5.2 : Major social risk affected by the vulnerable groups in disaster situation

Group or Sector	Risks and Vulnerabilities
Children: 0 - 5 years old 6 - 14 years old	Illness and death, hunger and malnutrition. Stunted growth, poor education quality
Youth: 15 – 24 years old	Low human capital, unemployment and underemployment, substance abuse
Elderly: 65 years old and above	Illness and death, low or lack of income
Women	Large family size, domestic violence, unemployment and underemployment
People with disabilities (PWDs)	Low human capital, unemployment and underemployment, lack of access to services
Indigenous People	Dislocation due to armed conflicts, natural disasters, development projects, lack of shelter and housing, access to water and sanitation
Rural and Urban Informal Communities	Dislocation due to armed conflicts, natural disasters and demolitions, lack of shelter and housing, access to water and sanitation

In the case of Kerala, the major vulnerable groups affected by the recent floods are children, youth, elderly and women. Perhaps, the children and the elderly are the most affected sections in the population as of any disaster. Most of the schools and hospitals in the affected regions were closed during the disaster period. Illness and death tolls rose slightly in the severe affected areas, but such information were not given public attention. The prevalence of illness has increased substantially. There are many families still living in shelters given by other people or voluntary organizations.

Social Protection Programmes

There are many studies which examined the functioning of social protection system in developing countries like India and Kerala in particular (Agarwal, 1990, Arjun S et al, 2006, Gopal 2006, John, 2004,

Kannan 2015, Nair R 2010, Rajan 2002, Sarkar 2004, and Tharamangalam 2011). Even though, these studies were not examined from the context of a disaster situation, it offers some insights regarding the evolution and prevailing condition of social protection in the study area.

While considering social protection, the usual thoughts coming to one's mind would be the government programmes implemented at the ground level. Most of the social protection measures in the country were initiated after 1970s. The food and nutrition programmes particularly the public distribution systems (TDPD), Mid-day meals, Integrated Child Development Schemes (ICDS) have made substantial impact on the poor in Kerala as of other states in the country. The housing programmes like IAY for rural areas, and the self-employment programmes like Swarna Jayanti Grameen Swarojgar Yogana (SGSY), Self-help groups, Prime Minister Employment Generation Programme (PMEGP), and Swarna Jayanti Shahri Rojgar Yojana (SJSRY) have a remarkable impact but not without its short-comings. Several employment programmes initiated since 1970s were later merged into NREGS programmes in rural areas.

In the case of Rashtriya Swasthya Bima Yogana (RSBY), all poor workers and their families are covered. This scheme is funded by central and state governments with a ratio of 75:25. The scheme provides cashless insurance cover upto Rs 30,000 per month for hospitalization. Migrant workers are also covered within the scheme using smart care. All pre-existing diseases are covered in the scheme. This programme has received encouraging response among the common people in the country.

Universal social security programmes were only recently initiated by the central government. In 2015, the central government has introduced a life insurance scheme named PM Jeevan Jyoti Yojana. Also, started an accident insurance scheme named PM Suraksha Bima Yojana, and a contributory pension scheme for unorganized sector workers named Atal Pension Yojana.

In Kerala, almost all social protection programmes and schemes are implemented through the Kerala Social Security Mission, a charitable society under Government of Kerala headed by the minister for social justice and health. The major schemes implemented by this agency at present are Cancer Suraksha, Snehapoorvam, Thalolam, Aswasakiranam, Hunger Free City, Vayomithram, Karunya Deposit, Prathyasha, Snehasanthwanam, and Samashwasam. The two programmes that received much attention recently would be the Disability certification camps and

Disability Census 2014-15. All these initiatives are for senior citizenes, children, differently abled, women and persons with chronic illness.

The Need for Reorientation

All the above mentioned central and state programme initiatives must be re-designed to suit the post disaster situation in the state. It is generally an accepted fact that in post-disaster situation, the social protection measures become shock responsive and therefore must focus on the following major areas to reduce the vulnerability conditions.

- Finance: Exploring whether and how the intervention might be funded
- Targeting: Considering who the intervention is designed to reach
- Data management: The merits and challenges of using the readymade databases
- Transfer values: Determining how much support to provide
- Payment/Distribution Mechanism: How the support will reach the target recipients
- Coordination: Linking all relevant sectors
- Communication: Discussing the intervention with Communities

In the case of finance, anticipation of potential funding requirement must be conserved seriously. Before you secure timely and predictable financing for shock responsive social protection, it is important to establish or strengthen systems for need analysis and funding requirements.

The channels of financing can be (i) Contingency funds: Funds that can be called upon in emergency must be accompanied with clear allocation of responsibilities for ensuring that they are adequately resourced and disbursable in a timely manner. (ii) Contingency credit lines: the release of emergency credit might offer liquidity in the aftermath of disaster. (iii) Disaster insurance: Insurance remains an alternative through which cost of responding to disasters can be smoothed

Targeting the intervention group is another aspect. A core principle of an effective shock response is to be clear about who needs assistance and what support they require. This means making good use of vulnerability assessments and need assessments. Authorities responsible for emergency response have processes for deciding where to go and whom to support. An important question then is whether the social protection sector can contribute as an effective alternative or complement to these processes from its experience with supporting chronically poor and vulnerable households. Further, it is also pertinent to recognize the advantage and challenges that might emerge in different contexts which needs anticipation of potential funding requirements.

Data management is crucial in planning. Government and intervening agencies often explore whether data bases that have been compiled for one primary objective such as storing data about the beneficiaries of a social assistance programme-can also be used in response to a crisis. The usability of these data bases is closely associated with the way that beneficiaries of the original programme have been registered and targeted. The five issues that needs consideration is (i) Relevance (Does it contain the right variables), (ii) Completeness (Does it contain data on all people/ households in the affected area or in need), (iii) Data currency (Is the information up to date), (iv) Accessibility (Can the information be accessible to people who need it) and (v) Accuracy (Is it free from errors).

Transfer Values determine the extent of support needs to provide to the target population.Humanitarian intervention aim to ensure that those affected by the crisis receive assistance that meets their needs. Subject to the availability of resources there has to be an inevitable trade-off between scale and value - supporting all those in need with less or providing adequately to fewer people. This calls for a coordination and harmonization of transfer values across the responses.There can be a vertical expansion which entails top-ups of the existing beneficiaries. A Horizontal expansion involves inclusion of new beneficiaries. Another alternative is the fresh component of assistance that is non-existent

Payment Distribution Modality is another important aspect. The System may vary from staff travelling to communities for distribution of physical cash or goods, to payment service providers such as banks/post-offices and electronic means of transfer to individual's bank accounts. Each combination of disbursement means has its implication for appropriateness accessibility, robustness and degree of integration. The challenges are (i) How best can you maintain the delivery of the routine social protection intervention in the event of a shock; (ii) How can you make sure that the emergency assistance is channeled through the same system which neither compounds the burden on the routine intervention nor suffers from the similar disruption.; (iii) If emergency assistance is delivered through a different system will it create any unacceptable difference between the support offered to routine social protection and emergency beneficiaries.

Another important aspect would be the proper coordination of all relevant sectors. Coordination is necessary for improving understanding of one another's fields, and strengthening engagement between programmes and its delivery system.

Strong communication with the community is important when humanitarian assistance is delivered through social protection programmes, particularly when there are multiple delivery channels. If social protection system and programmes are used for shock response then considerable attention needs to be paid to ensure improved coordination. Attention needs to be paid to improving understanding of one another's fields and strengthening engagement between programmes and its delivery system.

Conclusion

Even though we have many social protection schemes and programmes the overall expenditure on these schemes is very limited. Even in the country, the overall public expenditure on social protection excluding the public health care is only 1.5 percent of the country's total GDP. This share of expenditure is relatively lower when compared with many middle income countries of the world (ILO, 2015). Kerala state also faces scarcity in public funding for social protection schemes. It is high time to consider the various options for generating more funds for the purpose.

Many arguments are alive in the discussion on effective implementation of social welfare programmes in the state. The well-known model named, the residual view, usually takes relative advantage to a great extent in the context of the scarcity of public finance in the state. As per the residual view the social welfare services are to be provided only when individual's needs are not properly met through other social institutions, mainly the family. Social welfare schemes should not be provided until all other measures are completely exhausted. In the United States of America funds and services are provided for a short period and is withdrawn immediately after the individual or family regains their ability to manage themselves. During the disaster period the relief funds were mostly distributed in biased manner in many parts of the state. Most of the personnel in local government took a stand that some persons who are capable to manage themselves, need to be given government assistance. However, some other grama panchayats took a stand to provide universal assistance.

The institutional models are also relevant to some extent. Individual's problems arise out of the factors that are not within his/her control. The disaster came because of environmental factors. Individual efforts are not enough to overcome this disaster. Social institutions are required for environmental protection and building public infrastructure.

In Kerala context, it is not possible to organize the whole programmes of social insurance, residual welfare, personal social services and medical care solely by the local self-government or state government. Public private partnership would be considered as better option for better results in terms of both quantum and quality of schemes. For this, state government ideologies have to be moulded towards the interest of general public, and the emerging globalized socio-political environment. Non-government organizations which are very much involved in the welfare of marginalised sections of the community have to be part of the welfare programme initiatives.

In the implementation of programmes and schemes, the views of people with regard to fund utilization is worth considering. It has been argued that social protection funds do not reach the real target population. In reality, it is a waste of government funds. Very often there is no match between the risks and programmes. Many programmes are implemented without understanding the magnitude of the needs of the vulnerable population.

Among wage employment programmes, NREGA is the most effective one than any other previous programmes implemented in the state. This programme could empower women substantially in terms of improvement in their financial situation and building social networks. Even senior citizens, irrespective of gender, were able to transform their social status from a non-productive person to an economically active person. This programme has less corruption, and no contractor or middle man is involved in the execution of activities. Accountability is a must in all process of implementation. Social auditing is strict in the programme. One of the issues that came to light in Kerala is that most of the works are carried out for not building public infrastructure, but for activities in private land or properties. This is against the actual visualization of the programme. Very often, at the implementation phase, local self-governments could not find the useful public task for the NREGA works. In those places, the work is only with the intention to provide jobs to the local people. In the post disaster situation we should reconsider the current way of implementation of this programme towards building the public infrastructure and preserving the natural resources. In the areas unaffected by disaster, this programme may extend to agriculture sector particularly in rice cultivation and livestock keeping to ensure the food security of the state.

Finally, the state needs a revisit of its all social protection programmes and schemes. The programmmes must be redesigned according to the demand and needs of the population at the grass root level. Mere implementation of the programme as it is designed in the centre may not be effective. In fact, more dialogues and intellectual exercises are needed for reorientation and effective implementation of existing programmes to make the most of it. It is high time for the state government to shift their priority from their latest renaissance agenda to the core development and livelihood issues of the state.

References

1. Agarwal, B. (1990): "Social security and the family: Coping with seasonality and calamity in rural India", *The Journal of Peasant Studies*, *17*(3), 341-412.
2. Arjun Sengupta, KP Kannan, and Ravi Srivastava (2006): "A major national initiative", *Economic and political weekly*, Vol41, no32, August 12.
3. Arun, S., & Arun, T. G. (2001): "Gender issues in social security policy of developing countries: Lessons from the Kerala experience", *International Social Security Review*, *54*(4), 93-110.
4. Devika, J., & Thampi, B. V. (2007): "Between 'Empowerment' and 'Liberation'- The Kudumbashree Initiative in Kerala", *Indian Journal of Gender Studies*, *14*(1), 33-60.
5. Gopal, M. (2006): "Gender, ageing and social security", *Economic and Political Weekly*, 4477-4486.
6. International Labour Organization (2015): *World Social Protection Report 2014-15.*
7. John C.P (2004): "Social Security and Labour Welfare with Special Reference to Construction Workers in Kerala", *Discussion Paper No. 65*, Kerala Research Programme on Local Level Development, Centre for Development Studies, Thiruvananthapuram
8. Justino, P. (2007): "Social security in developing countries: Myth or necessity? Evidence from India", *Journal of International Development: The Journal of the Development Studies Association*, *19*(3), 367-382.
9. Kannan K.P (2015): "Corporate capitalism in the name of social security", *Economic and political weekly,* Vol. 50, No 12, March 21.
10. Kannan, K. P. (1995): "Declining incidence of rural poverty in Kerala", *Economic and Political Weekly*, 2651-2662.
11. Kannan, K. P. (1995): "Public intervention and poverty alleviation: A study of the declining incidence of rural poverty in Kerala", India. *Development and Change*, *26*(4), 701-728.

12. Kannan, K. P. (2007): "Social security in a globalizing world". *International Social Security Review*, *60*(2 3), 19-37.
13. Kannan, K.P & Vijayamohanan Pillai, N (2007): "Evolution of Social Security in the Lap of Public Action: Recounting the experience of Kerala", *MPRA_paper_ 9691*.pdf
14. Kannan, K.P & Vijayamohanan Pillai, N. (2007): "Social security in India: The long lane treaded and the longer road ahead towards universalization", *MPRA_paper_9601*.pdf
15. Kumar, S. V. (2003): "Economic security for the elderly in India: An overview", *Journal of Aging & Social Policy*, *15*(2-3), 45-65.
16. Minimol, M. C., & Makesh, K. G. (2012): "Empowering rural women in Kerala: A study on the role of Self Help Groups (SHGs)", *International Journal of Sociology and Anthropology*, *4*(9), 270-280.
17. Mohanan Pillai S (1996): "Social Security for workers in unorganized sector: Experience of Kerala", *Economic and Political Weekly*, Vol. 31, No. 31 (Aug. 3, 1996), pp. 2098-2099+2101-2107
18. Nair, R. (2010): "Public intervention for food security: the case of Kerala". *Journal of Social and Economic Development*, *12*(1), 43-71.
19. Prasad-Aleyamma, M. (2011): "A market place for migrants: Mobility, settlement and social protection in Kerala", *Migration and Social Protection* (pp. 164-182), Palgrave Macmillan, London.
20. Rajan, I. (2007): "Population ageing, health and social security in India", *Centre for Development Studies Discussion Paper*, *3*.
21. Rajan, S.I. (2001): "Social assistance for poor elderly: How effective?", *Economic and Political Weekly*, 613-617.
22. Rajan, S.I. (2002): "Social security for the unorganized sector in South Asia", *International Social Security Review*, *55*(4), 143-156.
23. Rajan, S.I., Perera, M., & Begum, S. (2002): "Economics of pensions and social security in South Asia: Special focus on India, Sri Lanka and Bangladesh", *Fourth Annual Conference," South Asia Network of Economic Research Institutes (SANEI), Dhaka, Bangladesh, August* (pp. 28-29).
24. Sarkar, S. (2004): "Extending social security coverage to the informal sector in India", *Social Change*, *34*(4), 122-130.
25. Tharamangalam, J. (2011): "Is Food Insecurity in Kerala a Myth?", *Economic and Political Weekly*, 69-71.

6

Common Property Resources and Environmental Justice in a World at Risk: Lessons from the 2018 Kerala Floods

Janki Andharia and
Tania Teresa Mathews

Introduction

The floods in Kerala between June and August 2018 dealt a colossal blow to the state, exposing the fragility of its ecosystem given the pre-existing diversity in its geography as well as the ongoing human interventions and processes that are considered 'markers' of its economic progress. As 12 of the 14 districts in the state contended with the floodwaters gushing from dams and supplemented by persistent and unprecedented rainfall that submerged or swept away entire homes and bridges, many parts were additionally devastated by a series of landslides[1]. Despite the enormous setback both in terms of human impact and economic cost, the emergency response and relief undertaken by the citizenry greatly supplemented the efforts of rescue and humanitarian assistance by personnel deployed in the region, garnering praise from all quarters.

However, the extent of damage also compel a serious reflection on causation and preparedness of the state to deal with an extreme weather event that Kerala experienced. The state also gathered criticism from many including the scientific community in terms of the increased vulnerability of many areas owing largely to inappropriate or unplanned land use keeping the idea of river basin managemnet in mind, excessive damming of rivers and accumulation of water in the reservoirs, the absence of an integrated hydrological plan (or its implementation) for water management around dams, and quarrying activities - all of which contributed towards an increased hazard risk. As the number of hydrological calamities, namely floods and accompanying landslides, constitute nearly half the average number of annual disasters, their frequent resurgence warrants urgent steps towards climate change adaptation (CCA) and improving the means of Disaster Risk Reduction (DRR). Kerala's experience provides valuable lessons not just in terms of the importance of community participation in the recovery process, but also highlights crucial lacunae in the implemntation of its development goals which ignored the potential ecological and economic damage that were only elucidated by the disaster.

The most adversely affected in a disaster are communities that are marginalized even in the pre-disaster scenario (such as the Muthuvans of Idduki or other tribes from Wayanad). Communities entrenched in poverty or with low means of income, following traditional livelihoods and subsistence agriculture are usually the most vulnerable groups. The reliance on their local environment for crucial resources makes them doubly disadvantaged and vulnerable in the event of a disaster. This paper argues that incorporating a human rights perspective in understanding and framing disaster response affirms the rights and dignity of vulnerable people and can provide opportunities to transform societies and deepen equality in the aftermath of disasters. Laying the foundation of the discourse on disasters and climate change, the paper then moves on to the importance of environmental justice and the governance of Common Property Resources (CPRs) which together work towards environment sustainability and engendering community participation. Finally, it highlights the lessons learnt from the recent floods in Kerala, both in terms of the success of participatory recovery, and the shortfalls in mitigating disaster risk. The recommendations that follow seek to provide at least one frame of looking at the future planning process with the idea of risk-informed and climate-sensitive planning in the state.

The Global Context: Moving from Short-Term Disaster Response Strategies to Long Term Climate Sensitive Development

As the second decade of the 21st century draws to a close, the world is contending with rising inequalities, socio-political conflict and the effects of environment degradation. Global extreme poverty has been decreasing, but the rate of this decline has decelerated; while increased privatization in a globalized, neoliberal world has significantly contributed to flourishing industry and rising standards of living for the middle classes. This has been at the cost of the environment and the rising precarity of the already marginalized - rural poor and tribal communities. With the majority of the natural resources being claimed by the state or privatized, the loss of access to the community 'commons' has restricted those engaged in traditional livelihoods, who are also the worst affected because of their dependence on the environment and its natural resources to sustain themselves.

Data suggests that in a world 'at risk', climate related disasters have come to dominate the risk landscape to an extent where they now account for more than 80 percent of all major international reported disasters, with hydrological disasters alone accounting for over 40 percent of them. According to the Centre for Research in Epidemiology of Disasters (CRED) in Belgium, more people are in harm's way, i.e., vulnerable, than they were 50 years ago. Building on flood plains, earthquake zones and other high-risk areas, has increased the likelihood that a routine natural hazard will become a catastrophe. Disasters force an estimated 26 million people into poverty every year. Climate change will further increase the magnitude and frequency of extreme events in the future. In the recent disaster in Kerala, official statistics indicate that 1.2 million people were displaced.

The World Atlas of Natural Disaster Risk of 2015, systematically analysed vulnerability to 11 natural hazards worldwide and calculated the risk they pose to lives and livelihoods, country by country (171 countries). The Atlas comprises over 300 maps of the risk of 11 types of natural hazards - earthquakes, volcanoes, landslides, floods, storm surges, sand-dust storms, tropical cyclones, heat wave, cold wave, drought and wildfire - as well as maps for multi-hazard risk. It also displays global hotspots and where a disaster is most likely to occur. The World Risk Index is calculated along 4 components:

1. **Exposure to natural hazards** such as earthquakes, hurricanes, flooding, drought and sea-level rise
2. **Vulnerability** as dependent on infrastructure, nutrition, living conditions and economic circumstances
3. **Coping capacities** as dependent on governance, preparedness and early warning measures, access to healthcare, social and material security.
4. **Adapting capacities** with respect to impending natural events, climate change and other challenges.

The Intergovernmental Panel on Climate Change (IPCC) in a Special Report (2018) on 'Global Warming of 1.5 degrees Celsius' has stated that the last 18 years have been the warmest since the documentation of the climate measurements began. The 1 degree Celsius increase above pre-industrial levels since the mid-1800s[2], points out the need to limit further increase in global temperature to a maximum of 1.5 degrees Celsius above the pre-industrial levels in order to keep global sea level rise 0.1 m lower by 2100. If left unchecked, global warming would cross the 1.5 degree Celsius increase, worsening the risks of heat-waves and drought, greater sea level rise, flooding and rainfall. Maintaining the temperature rise under 1.5 degrees could tremendously improve the condition of various delicate ecosystems like the coral reefs or the melting glaciers of the Arctic. Climate change adaptation (CCA) can only be achieved through paradigm shifts in the consumption and production processes of global resources - moving towards renewable sources of energy, constructing sustainable infrastructure and providing livelihood opportunities in the process. Urgent action alone can help reduce the risk of extreme climate events. The IPCC in its Special Report (2012) on Extreme Events (SREX) highlights the significance of vulnerability and exposure in manifesting extreme climate events into disasters. It also expounds on the impact of climate-induced disasters which include high-intensity hurricanes, extreme precipitation, forest fires, and heat-waves - posing increased risk to life and property. Extreme weather does not prove the existence of global warming, but climate change is likely to exaggerate it - by disturbing the flow of ocean currents, inducing hot and cold vortices forming tornadoes or extreme cold weather respectively, bolstering heat waves, lengthening droughts and causing more precipitation and flooding. Hurricanes or cyclones become more destructive as warmer ocean waters feed more energy to the storms;warmer air also carries more moisture for devastating rainfalls, while rising sea levels lead to more flooding. Geologists contend that we have already entered the era of Anthropocene - the sixth mass extinction on a planetary scale with human activity as the major cause for the first time (Ceballos, et. al., 2015; Barnosky et. al., 2011).

This destruction is driven by several factors: extinction rates of flora and fauna are far above the long-term average; at the present rate, about 75 percent of species would become extinct in the next few centuries if the current trends continue. Increased concentrations of carbon dioxide in the atmosphere by about 120 ppm since the industrial revolution because of fossil fuel-burning, leaving concentrations today at around 400 ppm and rising, is yet another factor driving this mass extinction. According to some research, we have had the largest impact on the nitrogen cycle in 2.5 billion years. Humans have dumped large amounts of plastic in all the water bodies and oceans that microplastic particles are now virtually ubiquitous. Plastics will leave identifiable fossil records for future generations to discover. With our heavy fertiliser use, the nitrogen and phosphorous in our soils has doubled in the past century.

Patterns of economic development, rising inequalities, and population growth are the most critical factors in explaining the disastrous impacts of the hazard events. Managing our relationship with the biosphere is critical - where the challenge is to balance global strategies with regional ones. What is that balance and how is to be established - are questions that form much of the debate around nature of development pursued and its global impact. With this information, how does the world prepare for disasters? The strategies for disaster risk reduction must be an integral component of comprehensive strategies for sustainable development. How will this be achieved?

There is a clear need to move beyond the generally widespread short-term approach to assessing disasters with a more global, political economy approach wherein aspects such as preparedness, protection of particularly vulnerable groups and risk management must be foregrounded in the context of both the analyses that are undertaken and future measures - not at a local scale but at a global level. The dynamic changes observed in development planning and international cooperation in recent years involve the adoption of new frameworks such as the SDG agenda, Sendai Framework for DRR, Agenda for Humanity, and Commitments to Accountability to Affected Populations (C-AAP) to name a few, and convergence across these frameworks and the various stakeholders at the national and international levels. Their focus on social, economic and environment sustainability, while 'leaving no one behind' necessitates an egalitarian approach to yield equitable outcomes at the grassroots. Balancing the two - the global and the local, however remains a fundamental challenge. And it is here that the idea of Environmental Justice and Common Property Rights become significant. Both are closely intertwined.

Historically, CPRs are those typically used by rural and/or traditional communities who depend on natural resource systems to sustain their livelihoods and have acquired or inherited skills and technical knowledge over generations of engaging with the natural resource system.

Common Property Resources and Environmental Justice

Environment degradation and loss of traditional livelihoods at the grassroots, or "micro-level" issues are cumulatively responsible for "macro-level" concerns regarding the changing global climate and the degradation and pollution of the environment and the loss of its biodiversity (Runge, 1990; Ostrom et al, 1999). Such a micro-level instance of the "tragedy of the commons" - originally coined by British economist William Forster Lloyd in his lectures on population growth in 1833, thereafter popularized in academic literature by ecologist Garret Hardin in his eponymously titled 1968 paper - refers to the over exploitation of common property resources or common pool resources (CPR) which constitute the set of natural or human-made resources in a system that are typically 'rivalrous' and 'non-excludable' - differentiating them from private (rivalrous and excludable) and public (non-rivalrous and non-excludable) goods. However, the non-excludability of these resources is limited to a localised community or group that typically accesses these resources for its consumption or livelihood, and whose norms govern the sustained and equitable access of all members to the resource - such as forests, fishing grounds and irrigation systems.

While Hardin (1968) was a landmark in the discourse on the use of common resources, and initially found much favour in academia, it assigns blame for the abuse of commons on their traditional users: the cumulative effect of promoting the individual user's self-interests in their appropriation of these rivalrous resources being the primary cause of its 'tragic' overuse - with the economic justification that the marginal utility or benefits derived by the user are significantly larger than any marginal cost borne by them. The arguments put forth in Hardin's paper and supporting literature thus legitimized the acquisition and management of common resources by non-commoner agents - chiefly the state and the private sector (Runge, 1990; Ostrom, et al, 1999). Runge cumulatively extends this notion of the tragic overuse of commons resources to the macro context - at the national or global level. However, unlike Hardin, he attributes the responsibility of the overexploitation of the commons not on the "commoners" but on the state and the role of private entities taking over the traditionally common properties of indigenous communities; empirical evidence and literature since the 1980s has criticised Hardin's thesis. Runge (1986; 1990) notes the case of the enclosure system in 15th and 16th century England and Western Europe and the consequent rise of private property systems which significantly declined the use of common pastures and woodlands that village communities across Europe had been dependent upon; additionally, Runge also cites examples of extensive deforestation in Latin American nations following state policies to acquire forest covered lands and the subsequent impact of privatization and the incapability of the state to manage the vast resources acquired. Rather than correcting the 'mismanagement' of the commons, the rise of state acquisition and private property further degrades the environment and widens the inequalities between classes. This pattern is also reflected across the globe and especially rampant in developing economies.

Damodaran (1991) elucidates the same in the Indian context with the state acquisition of forests since the 1860s and the segregation of state and community rights over the forest resources. The Indian Forest Act, 1865 subsequently replaced by the Forest Act of 1878 were legislations that sought to "eliminate the rights of people" dependent on these lands prior to their acquisition by the British colonial administration, particularly those rights clashing with state interests (ibid). These lands - declared 'reserved' and 'protected' as forests were the large wastelands adjoining *zamindari* estates and village communities and had been largely common or community-owned properties. An NSSO study (1999) conducted in the 54th Round corroborates this, with an enquiry on common property resources particularly land resources in rural communities, defining common property resources as "all such resources which are meant for common use of the villagers" which had been freely available to the rural population prior to the British colonization and administration in India. These include village pastures and grazing grounds, village forest and woodlots, protected and unclassed government forests, waste land, common threshing grounds, watershed drainage, ponds and tanks, rivers, rivulets, water reservoirs, canals and irrigation channels. The report notes that the "extension of state control over these resources" resulted in the decline of the indigenous management and governance systems, thereby substantially decreasing CPRs available to the rural population. Damodaran (1991) notes that the state justified the "constitution of common lands into forest lands… to mitigate the free-rider problem through exclusion of user rights". This 'free-rider' problem that gives rise to the tragedy of commons phenomenon is however evaded in the context of traditional communities that have in place complex norms that govern their livelihoods and dependence on the commons and practice sustainable methods of resource use (Runge, 1990).

Damodaran (1991) and Runge (1990) therefore call for the need to highlight this distinction to avoid confusing these 'free-rider' commons with ecologically and institutionally stable, common property resources. There is a fundamental distinction between these two types of common-access resources; in her seminal work, Ostrom (1990;1999) presents the complex and implicit governance structures and norms that constitute the management of common pool resource institutions - "natural and human-constructed resources in which (i) exclusion of beneficiaries through physical and institutional means is especially costly, and (ii) exploitation by one user reduces resource availability for others". Ostrom notes eight 'design principles' of common property resource management drawn from her observations of traditional communities' practices around the world (Ostrom, 1990; Cox, *et al*, 2010):

1. Well-defined boundaries: around the community of users and the resource system they utilize to ensure non-excludable access for all members of the group, but avoid the problems of open-access that 'free-rider' resources face (Damodaran, 1991).
2. Congruence between appropriation and provision rules and local conditions: i.e., proportional equivalence between benefits accrued to and costs borne by all members of the community thereby establishing equal rights and responsibilities.
3. Collective-choice arrangements: wherein community members utilize their comparative advantage through indigenous knowledge to devise the most efficient norms and guidelines on the management of the resource.
4. Monitoring: to critically track the compliance and effectiveness of the resource management practices and thereby make visible those who do not comply with the rules.
5. Graduated sanctions: the progression of incremental sanctions to appropriately deter participants from violation of community norms and maintain cohesion. Graduated sanctions also "maintain proportionality between the severity of violations and sanctions" (Cox, *et al*, 2010).
6. Conflict-resolution mechanisms: to complement the system of graduated sanctions and proportionality of benefits and costs, there is a need for an effective grievance redressal and conflict resolution mechanism especially given the rivalrous and non-excludable nature of common property resource systems.

7. Minimum recognition of rights: that "external government agencies do not challenge the right of local users to create their own institutions" or enforce their rules on the community's management of the resource system (Cox, *et al*, 2010).
8. Nested enterprises - i.e., that common property resource management units are arranged in a nested hierarchy - smaller common-property systems being nested in larger and still larger ones as in cases of pastoral and irrigation systems (ibid).

Thus, based on these principles put forth by Ostrom based on her observations of indigenous community management of resource systems, it may be understood that common property resources are effectively managed through people's engagement and collective action and therefore provide a suitable alternative or could complement privatization and state acquisition of these vast resources (Runge, 1986; Wade, 1986). Ostrom *et al* (1999) highlights a study by Sneath in 1998 of pastoralism in northern China, Mongolia, and southern Siberia, showing significantly larger degradation of pastures managed by state-owned agricultural collectives imposed in Russia and China and involve permanent settlements unlike in the Mongolian region where they are managed by traditional group-property institutions that involve large scale movements between seasonal pastures.

Common property resources in Kerala are largely within the domain of the coastal region and forests; the former concerned with the livelihoods of small fishing communities and their dependence on coastal waters and the lagoons. The continued survival of these fishing communities against large-scale industrial players is largely attributed to the collective action taken by the community members in preserving these common property resources they access. The latter is also significant considering the fact that nearly a third of the land in the state comprises forest cover. Kerala is home to a small and largely marginalized but significant tribal population, especially along the hills of the Western Ghats that border the state. The dependence of these tribal communities on the forests and other natural resources like rivers and ponds for their livelihoods is a controversial matter against the developmental goals of the state economy and the nation at large. Fortunately, with the aid of civil society interventions and initiatives of community members to promote dialogue on these contentious issues, tribal communities are now better equipped to secure their rights through legislation like the Panchayats (Extension to Scheduled Areas) Act, 1996 and Forest Rights Act, 2006.

In the context of Kerala, and for the country at large, the recognition of the right to common property resources and their management is a necessity for the welfare of communities that continue to sustain themselves on their traditional livelihoods as in the case of the small-scale fishing communities in the coastal parts and the tribal communities in the forest regions. The conflict of interests between the developmental priorities of the state that pander to the wants of private players in the market for the same resources against the welfare and needs of these inherently disadvantaged and marginalized groups necessitates civil society intervention and sustainable solutions that can enable their livelihoods to flourish, protect the environment, and prevent a 'tragedy of the commons' scenario.

Rights and Duties: A Totalizing Concern

Environmental justice can be understood as the "fair treatment and meaningful involvement of all people regardless of race, colour, national origin, or income with respect to the development, implementation and enforcement of environmental laws, regulations and policies" (USEPA). It necessitates that all individuals be equally protected from environmental and health hazards, thereby reducing vulnerabilities; and that they enjoy equal access to the decision-making process to have a healthy environment to live, learn, and work in. Chopra *et al* (1989), characterise participation with one or more of the following: groups of people; adherence to a set of norms; and involving "empowerment and adjustments" between individual and group behaviour. Citing Olson (1965), which notes that small size and homogenous groups are optimal for collective action to effectively manage resources, they note that common interests of the group might influence its behaviour - thereby making a case for local communities directly affected by and dependent on the resource system to hold the rights of access and management of the core or stock resource and its fringes.

For instance, as Harris (2016) illustrates,environmental justice demands that people affected by pollution have a say in how the pollution is reduced or prevented, and it requires that affected people be allowed to participate in determining who benefits, and how, from the policies instituted to correct past pollution. Environmental justice movements begin at the grassroots,constituted by the affected citizens, civil rights activists, and environmentalists collaborating to secure the rights of low-income and minority communities to clean and safe environment. Historically the movement was documented and coined as such in the United States, however, examples of such mobilisation and grassroot-level participation for the sake of the environment can be found across the globe. These communities are typically moral economies characterized by a “subsistence ethic” and a tradition of conservationism (Scott, 1976). Iconic examples in India alone include the forest conservation Chipko movement of April 1973 in Uttarakhand to prevent large scale deforestation; and the recent success of the Kadar tribal community in the Vazhachal Forest region in Kerala in acquiring Common Forest Resource rights and harnessing their indigenous knowledge of ethnobotany to support the Forest Department of the state in surveillance and management of the forest resources is an example of the Joint Forest Management System - a form of participatory management of common pool resources in India.

Environmental justice is interdisciplinary as it incorporates dynamic and complex scientific issues with equally complex issues contending with the history of socio-economic, political and cultural structures such as the distribution or accumulation of resources against race, economic class, and power. It arguably requires that the actual distribution of environmental benefits and burdens be fair. As Harris (2016) succinctly notes:

> Justice is about determining to whom rights are owed, and to whom associated duties should be assigned. It is about how much of the burdens of protecting those rights each actor with duties should bear. The just use of natural resources is even more pertinent in poor and marginalized communities where scarcities and pollution can be life-threatening.... Justice is often about how policies are decided - procedural justice - and what comes from decisions and actions - consequentialist justice. To realize procedural justice, people’s rights should be respected during decision-making, and individuals and communities affected by decisions should be allowed to participate in the decision-making process, ‘climate justice’ is a totalising concern, that creates new leverage for late industrialisers, requires a proactive strategy, within a limited temporal horizon, embedded within an all-encompassing and radically challenging epistemology. It is, however forcing broad-scale transformations in terms of political contestation.

While the justice of the outcome may be hard to assess, it should at least avoid, what Moral and Social Philosopher and academic, Brenda Almond calls, ‘striking inequalities’ (Harris, 2016). People tend to value outcomes that they perceive to be just; most often when they are perceived to benefit one's community, even if only in the long term. The Native Americans embrace the Seventh Generation Principle which is the essence of the sustainable development agenda - the choices and decision making with regard to the current use of resources is to be made keeping in mind the needs of seven generations into the future. Sustainability and environmental justice thus work in tandem. Common Property Resources are integral to the discourse on sustainable development since they enable local or community level initiatives and collective action to shape the efforts towards conservation of resource systems while also allowing the community members to benefit from the core resources and its fringes in an equitable manner. CPRs are typically used by rural and/or traditional communities who depend on natural resource systems to sustain their livelihoods and have acquired or inherited skills and technical knowledge over generations of engaging with the natural resource system. In the context of the sustainable development agenda, the recognition of CPR rights would enable these oft-marginalized communities to develop local solutions to improve their socio-economic well-being while simultaneously taking corrective steps to protect the environment.

The Human-Rights Based Approach to Disaster Risk Reduction

According to a UNISDR Input Report (2014), human rights-based approach to disaster risk reduction may have the potential to decrease a country's vulnerability to hazards. This approach focuses on three key elements: non-discrimination; participation; and accountability. Therefore, irrespective of several policies on natural disasters, if fundamental issues such as human rights protection and empowerment of local communities are not tackled, the effectiveness of disaster risk reduction strategies will be weighed down (da Costa and Pospiezna, 2014).

In the aftermath of the disaster events of 2004-05, chief among them being the Indian Ocean tsunami (2004), and Hurricanes Katrina and Wilma (2005) there was increased recognition that disaster response involves more than just delivery of humanitarian assistance. Critical issues pertaining to human rights that have emerged in past crises and during disaster management processes include discrimination in aid provision on the basis of gender, age, or ethnicity; enforced and often unsafe relocation and resettlement; sexual and gender-based violence; loss of documentation; and lack of access to legal protection, health services, shelter, clean and safe water, education, or sufficient monetary compensation, among others. Moreover, it was found that the longer the period of displacement, greater the risk of human rights violations, often the result of inadequate resources, or inappropriate policies, neglect or oversight (IASC, 2008).

A growing understanding of the need to respect, uphold and promote the human rights of those affected by natural disasters thus emerged from a series of missions and evaluations by the Representative of the UN Secretary General on the human rights of internally displaced persons which reflected a lack of awareness on the relevance of human rights norms in the context of natural disasters. The UN Inter-Agency Standing Committee (IASC) thus adopted Operational Guidelines on Human Rights and Natural Disasters in June, 2006, to facilitate guidance to national and international agencies engaged in humanitarian assistance so as to better enable them to protect the rights of disaster affected individuals.

The Guidelines thus sought to bridge the gap between humanitarian action in emergencies and the standards for protection of human rights in conflict situations, thereby providing specific guidelines on addressing human rights concerns arising from natural disasters. They are informed by provisions in relevant international human rights law such as the Universal Declaration of Human Rights, International Covenants on Civil and Political Rights and on Economic, Social and Cultural Rights, the Convention on the Elimination of All Forms of Discrimination against Women, and the Convention on the Rights of the Child. Additionally, the Operational Guidelines also draw from relevant regional human rights conventions and standards such as the Guiding Principles on Internal Displacement, the Sphere Humanitarian Charter, Minimum Standards in Disaster Response, the IFRC Code of Conduct and the IASC IDP Policy[3]. As a consequence, affected individuals can seek protection and assistance from their governments.

Lessons from Kerala Floods

As a coastal state with a varied topography of mountains, rivers, and back-waters, Kerala has a high dependence on climate sensitive sectors such as fisheries, agriculture, forests and plantations. The livelihoods of the poor and marginalized labour depend on these sectors. With a population of 3.3 crores, residing in a geography of 38,863 sq.kms (density of about 860 per sq. km), the state is affected by the winds, sea erosion, soil piping, (KSDMA, 2016) low productivity, compounded by quarrying, sand mining, reclamation of wetlands, and rapid changes in land-use and land-cover - both legal and illegal. The state's total area accounts for only about 1 percent of the country's total area, however it stretches along the Malabar coast for approximately 580 km, varying in width from approximately 30 to 120 km. It is bordered by the states of Karnataka in the north, Tamil Nadu to the east, and the Arabian Sea in south and west. The high population density, rising demand for housing, rapid expansion of infrastructure and a high density of dams on rivers without comprehensive spatial planning processes create conditions that make the state vulnerable to both disaster and climate change.

The state is one of the most socio-economically developed among other Indian states with a high literacy rate, HDI, and other welfare indicators. The floods in July-August 2018 are considered to be the worst in 100 years. Weeks of heavy rainfall caused dams to open their floodgates, triggered landslides in the mountains, and swamped the coastal regions of Kerala. The media reported loss of over 500 lives and evacuation of over a million people. Could better planning and preparedness have avoided disaster or at least minimized the grave consequences?

What Kerala Can Learn The audit report revealed that, in Kerala, dam-break analysis was not conducted in respect of any of the 61 dams in the state (1980). It is worth noting that there are 184 Flood Forecasting Stations in India. However, there is not a single flood forecasting station in 10 of the states including Kerala (Pandey and Sengupta, 2018). Furthermore, the drainage systems in residential areas are not adequate. Improper sewage treatment in the residential areas, results in the clogging of storm water drains, which lose their capacity to carry extra load of flood water. Also, the solid waste management in the state is substandard and epitomizes the common property crisis, begging the question - "who does the waste belong to?' Furthermore, the canals are severely polluted and filled with water hyacinth. In addition, unchecked civil constructions for different 'developmental' purposes also impedes the functioning of the drainage systems. This kind of absence of macro-level strategy and planning and its adequate implementation is one of the areas that the state needs to make progress in.

Kerala needs to recognize that construction of flood water drains is as important as construction of roads and other infrastructure while expanding a city or a residential area. This understanding also needs to devolve at the level of the Panchayats. Also, low lying areas should ideally be used as drainage basins and not for construction activities in the name of development. Flood plain wetlands need to be conserved at any cost since they are unique ecosystems and play an important role in the hydrology of the region. Moreover, unscientific and uncontrolled sand mining of the river beds caused the lowering of the river bottom drastically and the water spread area has been reduced considerably. As a result, the velocity of the flowing water increases manifold and reaches the lower regions immediately after a heavy rain. Understanding that the environmental destruction of the flood plain wetlands and hills in the catchment areas (rain fed areas) aggravated the present flood situation needs to inform future . The water holding capacity of the inland flood plains, wetlands and paddy fields have drastically reduced due to reclamation and blocked drainage systems. Also, the water holding capacity of the earth on the hills have reduced due to the uncontrolled quarrying and violations of the environmental law.

In order to tackle these issues, Kerala needs to focus on building macro level strategies for appropriate land use planning and effective management of resources in convergence with micro-level planning.

Dynamic reservoir operations strategy for individual reservoirs as well as combined strategy at river basin level is to be developed to enhance their utility. Plans for both monsoon and non-monsoon periods are needed to keep a check on its functioning. For flood control, the strategy/protocol/ operating procedure must look into the monsoon trends, time period etc. to increase effectiveness. Drought proofing at river basin level and ensuring e-flows shall be the priorities for reservoir operations during non-monsoon periods. Furthermore, in order to ensure implementation of strategies, participatory institutional mechanism is required.

Another interesting technique to enhance the resilience of the state to floods is through the practice of Kerala's indigenous activity - *pokkali* farming. It is a unique farming system that produce salt and flood resistant rice and shrimp alternated on the same farms in Kerala's tidal-water tracts. Pokkali rice is tolerant to soil salinity and submergence and thus is very suitable for climate adaptive agriculture. On high tide or river flooding, it holds its grain-bearing head upright above water for about ten days, while the plant itself bends over and collapses. This particular system, therefore, plays a great role in maintaining the ecosystem of farm tracts (Vijayan, 2017; Jena, 2016; Ranjith, Karunakaran and Sekhar, 2018). Thus, perhaps there is a need to promote this farming system to ensure agricultural resilience and ecological balance.

What Kerala Floods Offer The disaster response and recovery in Kerala was indeed commendable. The rescue operations led by the state officials along with the rescue efforts carried out by the army, navy and air force were timely and effective. The team of medical doctors and other professionals worked round the clock to serve and support the affected. Apart from this, Kerala demonstrated how community involvement and use of local capacities and volunteers can play a significant role in disaster response. The fisher-folk proactively used their indigenous knowledge and skills to save lives during the floods. Their grit and desire to reach out were indeed admirable and inspirational. Moving beyond the speed with which it responded to the crisis and how the state mobilised its own capacities, the state's ability to leverage its decentralised model of development planning, for long term recovery while ensuring accountability to the affected population is an approach that has been adopted and briefly mentioned which promises to be a unique governance innovation and is still unfolding as this paper goes into print.

The Accountability to Affected Populations, is an initiative taking shape in an effort of Government to build Kerala Back Better, as more resilient state with the participation and feedback from people affected by the flood. This initiative has created a system to collect information on the post-disaster needs and feedback of the affected communities taking into account gender and diversity of the communities, with a comprehensive set of common tools and protocols that map the needs sectorally. The system rolled out across the districts impacted by floods with the involvement of Kudumbashree and the Self-Help Group network of the Local Self Government Department (LSGDs). The system provides real time information to select line departments of the Government of Kerala, as well as the State Disaster Management Authority. This pioneering initiative in India is institutionalized within Kudumbashree, enabling a connection between ongoing development planning processes and humanitarian response and long term recovery. It is also expected to contribute to preparedness and resilience-building for future disasters, should they occur.

The geographic and population coverage is vast - 489 most affected Grama Panchayaths/Municipalities/Municipal Corporations, 1% of the affected populations (19,869 out of 5.5 million)[4] will be reached. They will provide feedback and deepen the idea of people-led recovery and reconstruction planning in Kerala. The 'feedback dash board' available to the Revenue Department, the LSGD, Kerala State Disaster Management Authority (KSDMA) along with district collectors, would ensure that by 2019-20, all the 489 Grama Panchayath plans are risk informed and will have included feedback from the affected people such that local risks are reduced and disaster risks, minimised. The State Planning Board and

Kerala Institute of Local Administration (KILA) are important partners[5]. The 13th working group notified for Gram Panchayat Development Plan planning via the Government Order of September 29, 2018 will be supported to develop climate sensitive plans and Gram Sabha's discussions on environmental decision making.

Conclusion

The Kerala floods of 2018 was devastating. The impact of the flood could have been considerably reduced had the state incorporated and implemented DRR strategies in development. Improper sewage and drainage management, unplanned and unscientific constructions, excessive quarrying, etc. resulted in increased damage that otherwise could have been avoided. Thus, the paper argues for an approach to DRR that places the community at the centre of DRM, strengthens preparedness for response, interconnects hazard risks, environment and land use and suggests much more work on strengthening institutional capabilities across all levels. This provides at least one frame of looking at the future planning process with the idea of risk informed and climate sensitive planning in the state within the well acknowledged "Kerala model of development".

Notes

1. According to the PDNA Report (October 2018) between June 1 and August 18, 2018, Kerala received cumulative rainfall that was 42% in excess of the normal average. The heaviest spell of rain was during 1-20 August, when the state received 771mm of rain. The torrential rains triggered several landslides and forced the release of excess water from 37 dams across the state, aggravating the flood impact. It affected 5.4 million people, displaced 1.4 million people, and took 433 lives (22 May 29 August 2018). The seven worst hit districts were- Alappuzha, Ernakulam, Idukki, Kottayam, Pathanamthitha, Thrissur and Wayanad. Nearly 341 landslides were reported from 10 districts. Idukki, the worst hit district, was ravaged by 143 landslides.
2. 'Rapid, unprecedented change needed to halt global warming - U.N.', Nina Chestney, Jane Chung. Reuters. October 8, 2018. https://www.reuters.com/article/us-climatechange-ipcc/rapid-unprecedented-change-needed-to-halt-global-warming-u-n-idUSKCN1MI022
3. IASC, Implementing the Collaborative Response to Situations of Internal Displacement, Guidance for UN Humanitaran and/or Resident Coordinators and Country Teams September 004 (hereinafter IASC IDP Policy)·
4. The Grama Panchayaths/Municipalities/ Municipal Corporations are reached through a phased approach:
    - 104 gram panchayaths of Alapphuza and Wayanad (Phase-1)

- 385 gram panchayaths of Kottayam, Pathanamthitta, Ernakulam, Idduki and Thrissur (Phase 2) Dec to Jan 2019
- 489 gram panchayaths of Alapphuza, Wayanad, Kottayam, Pathanamthitta, Ernakulam, Idduki and Thrissur (Feb March 2019)

5. In addition UNICEF, the JTSDS of the Tata Institute of Social Sciences, Riddhi Foundation, Communicating with Disaster Affected Communities-Network (CDAC-N) and Sphere India are playing a supportive role until all functions are institutionalised.

References

1. Barnosky, A. D., Matzke, N., Tomiya, S., Wogan, G., O., Swartz, B., Quental, T.B., Ferrer, E.A. (2011): "Has the Earth's sixth mass extinction already arrived?" *Nature*, 471(7336), 51-57. https://www.ncbi.nlm.nih.gov/pubmed/21368823
2. Ceballos, Gerardo, et. al. (2015): "Accelerated Modern Human-Induced Species Losses: Entering the Sixth Mass Extinction". *Science Advances*, American Association for the Advancement of Science. 1 June 2015. https://www.advances.sciencemag.org/content/1/5/e1400253.full.
3. Chopra, K., Kadekodi, G.K., Murty, M.N. (1989): "People's Participation and Common Property Resources", *Economic and Political Weekly,* pp. A189-A195.
4. da Costa, K, & Pospieszna, P. (2014): "Finding the missing thread: the inclusion of a human rights-based approach in tackling climate change mitigation, adaptation and disaster risk reduction". https://www.unisdr.org/we/inform/publications/49949
5. Cox, M., Arnold, G. and Tomas, S.T. (2010): "A Review of Design Principles for Community-based Natural Resource Management", *Ecology and Society*, Volume 15(4): 38.
6. Damodaran, A. (1991): "Tragedy of the Commons and Comedy of the Common Property Resources", *Economic and Political Weekly,* September 21, 1991. pp. 2213-2215.
7. Field, C. B., V. Barros T.F. Stocker, D. Qin, D.J. Dokken, K.L. Ebi, M.D. Mastrandrea, K.J. Mach, G.K. Plattner, S. K. Allen, M. Tignor, and P.M. Midgley (Eds.) (2012): *Managing the risks of extreme events and disasters to advance climate change adaptation,* New York: Cambridge University Press.
8. Hardin, G. (1968): "The Tragedy of the Commons", *Science*. 162(3859): pp.1243-1248.
9. Harris, P.G. (2016): "Environmental justice: Vitally important, but difficult to define". Essay drawing from Harris, P.G. (2016): *Global Ethics and Climate Change*, Edinburgh: Edinburgh University Press and works cited therein. March 8, 2016. Retrieved from: https://mahb.stanford.edu/wp-content/uploads/2016/03 MAHBBlog_EnvironmentalJustice_PHarris_ Mar2016.pdf
10. IASC (2008): *"Human Rights and Natural Disasters: Operational Guidelines and Field Manual on Human Rights Protection in Situations of Natural Disaster"*. Inter-Agency Standing Committee. March 2008. https://interagencystandingcommittee.org/system/files/legacy_fileshuman_rights_

and_natural_disasters_operationalguidelines_and_field_manual_on_human_rights_protection_in_situations_of_natural_disasters.pdf

11. Jena, Manipadma (2017): "Climate-Smart Pokkali Farming Needs Machines", *India Climate Dialogue.* https://www.indiaclimate dialogue.net/2017/10/16/mechanise-climate-smart-farming-food-security/
12. KSDMA (2016): Kerala State Disaster Management Plan Available at http://sdma.kerala.gov.in/wp-content/uploads/2018/11/Kerala%20State%20Disaster%20Management%20Plan%202016.pdf
13. Lloyd, W.F. (1833): *Two Lectures on the Checks to Population*.
14. MoE&F (2011): *Report of the Western Ghats Ecology Expert Panel,* Ministry of Environment and Forest (Rep), Government of India.
15. MoE&F (2013): *Report of the High-Level Working Group on Western Ghats,* Ministry of Environment and Forest (Rep). Government of India.
16. NSSO (1999): "Common Property Resources in India", NSS 54th Round. January - June 1998. National Sample Survey Organisation, Department of Statistics and Programme Implementation, Government of India. December 1999.
17. Olson, M. (1965): *The Logic of Collective Action: Public Goods and the Theory of Groups,* Harvard University Press.
18. Ostrom E., *et al* (1999): "Revisiting the Commons: Local Lessons, Global Challenges", *Science*. Volume 284. May 1999. pp. 278-282.
19. Ostrom, E. (1990): *Governing the Commons: The Evolution of Institutions for Collective Action*. Cambridge, UK: Cambridge University Press.
20. Pandey, K., and Sengupta, R. (2018): Economic and environmental aspects of Pokkali Rice-Prawn production system in central Kerala". *International Journal of Fisheries and Aquatic Studies*. 08-13. https://www.fisheriesjournal.com/archives/2018/vol6issue4/PartA/6-3-48-108.pdf
21. Runge, C.F. (1986): "Common Property and Collective Action in Economic Development", *World Development*, Vol 14, No.5, pp. 623-635.
22. Runge, C.F. (1990): "Common Property Resources in a Global Context", *Staff Papers Series,* Department of Agricultural and Applied Economics, University of Minnesota.
23. Sajeev, T.V., and Alex, C. J. (2017): "Mapping of Granite Quarries in Kerala, India: A ctitical mapping initiative". https://www.researchgate.net/publication/327622074
24. Scott, J.C. (1976): *The Moral Economy of the Peasant: Rebellion and Subsistence in Southeast Asia*, Yale University Press, 1979.
25. USEPA (n.d.) "Environmental Justice", United States Environment Protection Agency. https://www.epa.gov/environmentaljustice.
26. Vijayan, Roshni. (2016): "Pokkali Rice Cultivation in Kerala", *Agricultural Update*, vol. 11, no. 3. July 2016, pp. 329-333.
27. Wade, R. (1986): "The Management of Common Property Resources: Collective Action as an Alternative to Privatization or State Regulation", *Discussion Paper*, Research Unit, Agricultural and Rural Development Department, Operation Policy Staff, World Bank. May 1986.

7

Empowering Local Self Governments for Mainstream Biodiversity in Sustainable Development of Kerala

N. Anil Kumar and Suma Vishnudas

Biodiversity is life, and it is the most dynamic element of a bioregion for food and agriculture production. The Convention on Biological Diversity (CBD) came into force in the year 1992 signed by 168 countries and agreed upon the conservation of biological diversity considering it as a global asset. Conservation, sustainable use of the components of biological diversity, and the fair and equitable sharing of benefits from the use of genetic recourses are the objectives of the CBD. The convention recognizes, biological diversity as *our need for food security, medicine, fresh air, water, shelter and a clean and healthy environment* (www.cbd.int/intro/default.shtml).

Biodiversity has a critical role to play in dealing with two principal issues that limit social progress - climate resilience and under nutrition. According to FAO, since 1900s, some 75 % of plant genetic diversity has been lost as farmers worldwide have left their multiple local varieties for genetically uniform, high yielding varieties (FAO, 1999). Hence, it is important to place high on the development agenda, leveraging biodiversity for overcoming the climate related biotic and abiotic stresses, as well as bringing nutrition in food production and consumption, and thereby addressing the issues of climate resilience and under nutrition.

The second goal of 2030 U. N. Sustainable Development Goals is to "End hunger, achieve food security, improve nutritional status and promote sustainable agriculture". One of the targets under goal 2 is to "ensure maintenance of the genetic diversity of seeds, cultivated plants and farmed and domesticated animals and their related wild species through soundly managed diversified seed and plant banks at the national regional and international levels by 2020". Another target is to promote climate resilient agricultural systems all over the globe. Realizing the Sustainable Development Goals requires changes in institutional environments, information delivery and governance. Investments in biodiversity conservation need to be relooked. Above all capacity development at various levels is very important to achieve the targets.

Biodiversity mainstreaming is a long term iterative process of getting biodiversity concerns fully reflected in the policies, plans, and activities of the governance systems at local, sectoral or national levels in order to achieve the outcomes that concomitantly benefit biodiversity and development. It is as much a political and institutional issue as it is a technical one, requiring a process of changing mindsets to policy frameworks and practices. After the Second Global Plan of Action for Plant Genetic Resources (PGRs) for food and agriculture and several of the decisions of the CBD-CoPdecisions, necessary policies and measures were put in place by the Government of India to promote conservation and sustainable use of country's biodiversity. All the countries who were party to in the agreement of the Convention of Biological Diversity (CBD) have developed national strategies for conservation and sustainable use of biodiversity. Some specific measures undertaken by India to motivate local communities and institutions engaged in biodiversity management were enacting two national level legislations- Protection of Plant Varieties and Farmers Rights Act 2004 and Biodiversity Act 2002.India developed its National Biodiversity Action Plan in the year 2009 and that identifies the constraints and threats pertaining to biodiversity conservation (http://envfor.nic.in/division/national-biodiversity-action-plan nbab). India has developed a systematic plan to achieve this target by a proper institutional structure starting from MoEF&CC, Planning commission, National Biodiversity Authority, State Biodiversity Boards and Biodiversity Management Committees at Panchayath levels.

The Biodiversity Act & Rules mandate establishment of Biodiversity Fund at national, state and local levels for "the management and conservation of heritage sites; compensating or rehabilitating people from such designated heritage sites; conservation and promotion of biological resources; and most importantly "socio-economic development of areas from where such biological resources or knowledge accessed...." The National Biodiversity Authority, State Biodiversity Boards and the local level Biodiversity Management Committees are the institutions designated under the Biodiversity Act 2002 & Rules 2004 to advise/facilitate respective governments on all matters related to the biodiversity management. The prime responsibility at present for them is to help the governments to achieve the 12 National Biodiversity Targets and thereby contributing to achieve the Global Biodiversity Strategy Plan and Aceh Biodiversity Targets by 2020.

This paper discusses this context and the efforts of M.S. Swaminathan Research Foundation (MSSRF) in helping the elected Local Self Governments of a few Grama Panchayaths of Wayanad district of Kerala for mainstreaming the biodiversity concerns in the local development. This paper builds upon the idea of multi-level approach and processes needed from people to the policy makers for mainstreaming biodiversity. The gaps in awareness and institutional structures at different levels to achieve this goal are discussed in this paper. The need for consistent efforts to inculcate biodiversity conservation and sustainable use as a developmental agenda in the minds of elected representative, rather than an alternative idea away from the routine development processes is emphasized here.

Mainstreaming Biodiversity in Development

The Biodiversity conservation programme in the state needs to be oriented towards helping the elected Local Self Governments to mainstream biodiversity in local development and thereby achieving the various conservation goals. So, it is very important to discuss and identify the priority areas for the Biodiversity Boards vis-à-vis the 2020 National Biodiversity Targets. The Biodiversity Board of Kerala state, which started functioning in the year 2004 is considered as one of the best performing boards of the country. However, some of the important questions that needs to be addressed by the State Biodiversity Board are: Whether this body has to promote integrated conservation where *ex-situ, in-situ* and on-farm methods are combined and benefitting conservation and livelihoods simultaneously? How do we engage the local communities in the priority setting of biodiversity management? Will the preparation and maintenance of People's Biodiversity Registers be adequate? What are the ways and means for empowering the Biodiversity Management Committees and integrating the PBR data and concerns in the local governance? These questions have to be answered in the context of the 2030 Sustainable Development Agenda.

Kerala as a model for decentralized governance and development in the country had set the goal of sustainable development as early as 1990s. During the peoples' planning and decentralization, the core idea was to promote primary production systems and sustainable utilization of natural resources. Now Kerala is going through a critical development path, struggling to balance between the increasing individual needs of a developed population and arresting the depletion of natural resources.

The 2018 Big flood in the state aggravated the existing maladies, worsened the life struggles and vulnerabilities of the poor. One of the heavily affected districts of the state was Wayanad district- a critically important biodiversity terrain of Western Ghats. About 47 landslides, more than a thousand land slips and 234 land subsidence were reported. The impact assessment reports from the district revealed nearly 70% of the people and as much percentage of the production landscapes of this predominantly farm/plantation district have been directly affected. The heavy rainfall and the resultant flood, landslides and other calamities of severe magnitude had destroyed the ecological foundations of food and agriculture production of this district.

The Flood has created new kind of challenges in environment and development management. Climate change mitigation and disaster preparedness has become the two core urgent concerns in development. Natural disasters are uncertain and can create setbacks in social progress and long term impacts on economic development. Rebuilding process requires more extraction of natural resources and biodiversity in general. Reducing the direct pressures on biodiversity and promoting sustainable use should be the core of the future development plans. Also improving the ecosystem management and functions is vital. Functional and efficient ecosystems can only contribute to climate change mitigation and adaptation measures.

Climate resilient agriculture production landscapes and ensuring sustainable income from food and agriculture production are two such challenges for the state to ensure an equitable development. Improving resilience of food production landscapes through ecological intensification and maintenance of biodiversity in production system backed by a sustainable consumption plan could be the future strategy for the state. The first step towards achieving these goals is to include biodiversity on the agenda of development policy, planning and investment process and involving all stakeholders in the decision making process. More importantly, enhancing the capacity of local self-governments in mainstreaming biodiversity is the key to achieve the targets.An Action Framework with six clearly defined and developed mechanisms needed to achieve these outcomes are given below (see table 7.1).

Table 7.1: Action Framework for mainstreaming biodiversity in local governance

Attribute	Approach Mechanisms
Outcomes	Sustainable agriculture and food production, Fisheries, Forestry, & Tourism
Institutional Arrangements	NBA-SBB-BMC; SPB –BD mainstreaming M&E Group
Approach & Tools	Case studies; PBRs & BD Gramasabhas; BD based Enterprises Development
Capacity Building	Officials of the 4 sectors above; BMCs and community Leaders
Communications	Biodiversity Fair; BMC Websites; Social Media; Community Radio, people's biodiversity monitoring platforms.
Finance	Internal (Govt) Gene Fund, SDG Fund; External (donors-CSR) (5% of the plan fund of all LSGDs should be set apart for conservation, integration of biodiversity in all sectors of development)
Monitoring & Evaluation	SPB –SBB- BD mainstreaming M&E Group

MSSRF's Efforts to Mainstream Biodiversity and to Strengthen the Management

MSSRF, which played a key role in drafting the National Biodiversity Act 2002 and Protection of Plant Varieties and Farmer Rights Act in 2004 has established a Community Agrobiodiversity Centre in 1997 in Wayanad district. The Centre is engaged in activities of agro-biodiversity management by strengthening the systems in place as part of the emerged legislations, and reviving and revisiting the policies in the framework of emerging global sustainability frameworks. Some of the key activities are described here.

Peoples' Biodiversity Registers

The Biological Diversity Act 2001 directs preparation of People's Biodiversity Register' at every Panchayath, which documents knowledge of people on the biological diversity present in the Panchayath. The responsibility of preparation of such registers lies with the Biodiversity Management Committee. It should be maintained as a base document of all future conservation activities. MSSRF has taken leadership in preparation of this kind of a knowledge register, even much before the Biodiversity Act came in place in Kottathara, Meppady and Pozhuthana Grama Panchayaths of Wayanad. Since the implementation of Biodiversity Act, MSSRF has played a major role in preparingPeople's Biodiversity Registers. In 2013, Wayanad district became the first district in the state to complete preparation of PBRs in all its Panchayaths. The Community Agro-biodiversity Centre is actively engaged in assisting the three tier system of biodiversity action plan -National Bio-diversity Authority, State Bio-diversity Board and Biodiversity Management Committees. Though the state biodiversity board and BMCs have been constituted, the know-how on the application of Biodiversity Act is far too away from what is anticipated. Efforts at Panchayath level to facilitate the BMCs to undertake any activities other than drafting PBRs were very minimum. MSSRF focused on creating awareness for the BMC members on the Biodiversity & PPVFRA Acts and setting up of models at Panchayath level to facilitate conservation action under the leadership of BMCs.

Empowering the BMCs

During the years 2013-14, we conducted a study in 23 Panchayaths on the status of the implementation of Biodiversity Act in the district. The criterion set for assessment are awareness level of BMC members, Panchayath members and people on the Biodiversity Act, and use of Peoples Biodiversity Registers (PBR). The results of the participatory assessment shown that majority of Panchayath presidents were unaware of their BMC Chairmanship and so was the case of BMC Members about their BMC membership. Though People Biodiversity Registers (PBR) was prepared in most of the Grama Panchayaths, most of the BMCs did not conduct any follow-up meetings after the preparation of PBR. The agro biodiversity component is reviewed as very meagre in over 60% of the registers. So the decision was to expand the sensitization programmes on BD Act and powers of BMC. In the year 2014 only one Panchayath in the district has planned and implemented any single activity under BMC. The study could also identify that there is no mechanism to coordinate the activities of BMCs at district level. The BMCs were not represented in any of the planning bodies at Panchayath or district level and no working group and funding support was allocated at Panchayath level. In nutshell, BMC existed as paper institutions.

This study report was discussed at state level and generated discussion on how to improve the situation on biodiversity governance. The next effort was to build awareness among BMCs on BD Act and to set up some models in selected Panchayaths to formulate biodiversity conservation projects for BMC. Series of trainings, interactions and dialogues resulted in the formulation of four projects in Wayanad: (i) A community Seed Bank for rice in Vellamunda Panchayath, (ii) Regeneration of riverine ecosystem in Pulpally Panchayath, (iii) Restoration of riverine vegetation and conservation of Heronry in Panamaram Panchayath and (iv) a land use study in Vythiri Panchayath for developing land use guidelines. BMCs of the respective Panchayaths came forward by developing proposals and initiating activities at different levels.

Wayanad Community Seed Festival

Wayanad community seed festival was initiated by M.S. Swaminathan Research Foundation Seed Care and Wayanad District Tribal Development Action Council (WTDC), the two grass root institutions promoted by MSSRF in the year 2014. This 3-day event is an on going activity where a multi stakeholder participatory engagement ensured to democratize the ton-farm conservation. As part of building custodian farmer networks the details of farmers who are conserving native crop varieties in the district were documented.

Annual Seed festivals offer opportunity for the traditional farmers to organize, exhibit and exchange their seeds. The farmers are organized under the BMCs and represented respective Panchayaths in the fest.

Seed festivals attract people from different walks of life including children and it act as a knowledge sharing platform on agro-biodiversity. Many number of awareness classes are being conducted as part of the seed festival as a trans-disciplinary engagement and to have common understanding on the issues related to agro-biodiversity conservation and utilization. The annual seminars at the Festival discuss the policy gaps in integrating the agro-biodiversity concerns in to local development agenda.

Panchayath Level Initiatives to Conserve Agro-biodiversity

Three Grama Panchayaths of the district initiated programmes to set up Community Seed Banks. Edavaka Panchayath has initiated implementation of gene bank programme for tuber crops conservation since 2014. Edavaka has widened their conservation efforts to rice seed bank, special financial support to all farmers who cultivate traditional paddy varieties, medicinal plants and trees in the farm land. Vellamunda panchayath has initiated rice Seed Bank and Vythiri Panchayath initiated conservation plots for pepper varieties. Eight Panchayaths submitted projects for conserving different crop varieties in the last three years. Many of these projects get rejected in the processes of vetting or at the time of implementation due to conflicting clauses in the LSGD guidelines. It was also noticed that the officials at the vetting and implementation end are not sensitized enough on the changing developmental needs and concepts. Creating an economic benefit in conservation is the core issue in promoting sustainable use of biodiversity. The concept of "do ecology" which revolves around creating economic stake in conservation to serve simultaneously conservation as well as reduction of poverty becomes very relevant in this context. This is an opportunity for the state and panchayaths to create legally binding access to genetic resources for food, health and agriculture which can lead into development of "bio-enterprises" where benefits can be monetary with the knowledge providers.

This is critically important with respect to the conservation of medicinal plants that have use value amongst the tribe community of the district and for which market can stimulate their conservation and extensive cultivation. The CBD enabled Access & Benefit Sharing mechanism offers ways and means to help the individuals and communities to target community level market development for biodiversity.

Panchayath Level Seed Festivals

The BMC members of the Panchayaths who regularly attended the Community Seed Fest gradually started replicating the process at Panchayath level seed festivals. Such small localized seed festivals spread the message of conservation and use of agro-biodiversity to more number of people. BMC took major role in organizing such seed festivals and gradually identified their role in biodiversity governance.

Traditional Farmers' Directory

Wayanad community seed festival brought out a platform for many famers to interact with each other and exchange knowledge and seeds from different locations. These interactions led to the compilation of information about the traditional farmers of the district in the form of a directory. This was essential to create a network of farmers who conserve native varieties of crops and protect the seeds. The compilation helped in making a single point source on the details of conserved varieties as well as the farmers who can act as "conservation hubs" locally for protection of ago-biodiversity.

Recognition of Custodian Farmers and Community Agro-biodiversity Awards

MSSRF facilitated formation of a grass root organization among the custodian farmers of Malabar called 'Seed Care' in the year 2011. The objectives of this organization are to strengthen the conservation efforts of custodian farmers by building social networks, to work for protecting the farmer rights on the genetic resources they conserved, and to find markets for the value added native crop diversity and its products. They have initiated registration of 21 traditional rice varieties of Wayand as farmer's varieties under PPVFR Act 2001in the year 2012.

Wayanad Tribal Development Action Council (WTDC) is another Grass root organization formed exclusively among Adivasi communities of Wayanad to address the special developmental needs based on community biodiversity management in the year 2006. WTDC aims at eradicating poverty and malnutrition among Adivasis of Wayanad, protection of community rights on biological resources and traditional knowledge along with enhancing livelihood options. WTDC has got Genome Saviors' community Award from Government of India under PPVFR Act in the year 2012. They received an award of ten lakh rupees in recognition of the rice conservation activities of Mullu Kuruma and Kurichya communities of Wayanad.

WTDC has institutionalized two annual Community Agro biodiversity Awards for on farm conservation efforts of Adivasi farmers of the district utilizing the Genome Savior Award money. For the last five years WTDC is giving awards of Rs. 25,000 for the selected Adivasi farmer who conserve the most number of rice varieties in the district. Another Award of Rs.15000 is given for the contribution of agro-ecosystems conservation. These awards are distributed in the annual seed festival and given to the family including male, female and child members, in contrary to the conventional approach where often the male members are only recognized in public on behalf of the family.

Policy Advocacy

MSSRF submitted a set of policy recommendations regarding the agro-biodiversity management and BMCs after its five-year long interaction with BMCs and the Panchayath. The recommendations included year wise allocation of money for biodiversity management at Panchayath level, separate working groups for biodiversity, district level coordination of BMCs, representation of BMCs in District Planning

Committee and appointment of an officer in charge at district level to coordinate and vet proposals from BMCs.

During the current year, the LSG Department of Kerala has revised the SGD guidelines adding a new working group in Panchayaths to plan for biodiversity, environment and disaster management with 2% compulsory budget provision. This working group needs to be strengthened with inputs like basic data on biodiversity and natural resources of corresponding GP and the global situation in climate variations to plan future course of action. This is a welcome step and allocation of 2% fund for various activities under the working group can bring notable changes at grassroots level for biodiversity conservation and climate change mitigation.

Promoting the Wayanad District as a Carbon Neutral District

In Wayanad, farmers practice mixed crop cultivation in dry lands with many trees in the farm. Growing trees itself, is a conservation action in the context of climate change. However, farmers rarely benefit from this action till it's recognized by the state and value it for its contribution in combating climate change. Meenangadi Panchayath in Wayanad is piloting the carbon neutral project in the country through various efforts with the technical support of MSSRF.

Underpinning the Gains in Shade Coffee Production

Wayanad is the largest coffee growing area in the state. Almost 80% of the coffee production of the state comes from Wayanad. However, in recent times, the production has come down due to market fluctuations. Wayanad coffee plantations, are historically known for integration with abundant tree stocks and biodiversity; for example, some large farms have been identified with more than 40 different forest tree species, at least 8 rare and threatened frog species : Ground Frog (1sp*)* Bush Frog (5spp) and tree frog (2 spp) and over 100 bird species, more than in an equivalent area of farmland, and exceeded only by primary tropical forest. Recent research (Anil Kumar et al 2019, Karthika and Vishnudas, 2018) shows that Shade coffee gardens are almost equal to natural forest in case of certain faunal elements. This is a good evidence for marketing speciality coffee highlighting as grown in bio-diverse environment. Planting more trees in addition to existing trees will improve the carbon sequestration process combating global climate change. Coffee farms also check soil erosion and ensures nutrient cycling with deep-rooted trees in the farms ensuring sustainability. Shade grown coffee system amongst the small growers - marketing the coffee as a speciality product, carbon neutral and grown in bio-diverse environment, can improve the farm income and enhance the agro-ecosystem services. MSSRF advocates for shade coffee plantation as a climate resilient farming practice for Wayanad.

Conclusion

Kerala is going through the processes of rebuilding the state in the aftermath of the big flood. The UN report on this disaster and the rebuilding plan highlights the need for rejuvenating natural systems and bringing the concept of sustainability in practice. A key question however emerges: what are the evidences for those management practices, which poise protection and enhancement of human well-being, as well as production and production capability of biodiversity and ecosystem services for the new state? This question underpins sustainable development and it helped MSSRF to develop a conceptual model called 'C4 continuum' that pays concurrent attention to four fundamental purposes in the biodiversity sector, such as *Conservation, Cultivation, Consumption* and *Commerce* of genetic resources and ecosystem services management. This model produced several relevant and easily understandable indicators that are successfully used in measuring the biodiversity mainstreaming efforts in Wayanad district. The State Biodiversity Board in alliance with the National Biodiversity Authority and the Biodiversity Management Committees at the grass root level and the key stakeholder groups can take leadership to take this learning forward for the sustainable management of biodiversity sector of Kerala.

References

1. Nadesa Panicker Anil Kumar, Amsad Ibrahim Khan Saleem Khan and Vaniyan Balakrishnan 2019: Coffee, Climate and Biodiversity: Understanding the Carbon Stocks of the Shade Coffee Production System of India. In Handbook of Climate Change and Biodiversity, Climate Change Management, https://doi.org/10.1007/978-3-319-98681-4_7© Springer Nature Switzerland AG 2019 W. Leal Filho et al. (eds.)
2. Nitiayog, India (2017): *Millennium Development Goals Final Country Report,* Government of India, http://www.mospi.gov.in/sites/default/files/publication_reports/MDG_Final_Country_report_of_India_27nov17.pdf
3. Nitiayog, India (2018): SDG India Index, Baseline Report, 2018. http://niti.gov.in/writereaddata/files/SDX_Index_India 21.12.2018.pdf
4. UNDP (2018): Kerala Post Disaster Needs Assessment: Floods and Landslides August 2018. Executive Summary, UNDP. https://www.undp.org/content/undp/

en/home/librarypage/crisis-prevention-and-recovery/post-disaster-needs-assessment—kerala.html

5. FAO (1999): Food insecurity: When people must live with hunger and fear starvation: The State of Food Insecurity in the world. 1999. http://www.fao.org/NEWS/1999/img/SOFI99-E.PDF
6. FAO (2009): The international treaty on plant genetic resources for food and agriculture. Conservation and sustainable use under the international treaty. http://www.fao.org/3/a-i2579e.pdf
7. Karthika C. &Vishnudas C. K. (2018): "A comparative study of mixed-species bird flocks in shaded coffee plantation and natural forest in Wayanad, Kerala", *Indian BIRDS*, 14 (4): 97–102.
8. United Nations (2015): Transforming our world: The 2030 agenda for sustainable Development, A/RES/70/1, United Nations, https://sustainabledevelopment.un.org/post2015/transformingourworld
9. NBA, (2009): National Biodiversity Action Plan. http://envfor.nic.in/division/national-biodiversity-action-plan-nbap

Part - II

Rebuilding Kerala : Post Flood Challenges

8

Post-Flood Vision of a New Kerala

C.P. Rajendran and Kusala Rajendran

Introduction

A series of severe floods ravaged Kerala in July and August 2018, chiefly due to unusually high rainfall from 1 June 2018 to 19 August 2018. This was about 42% higher than the normal. Considered to be the worst flood disaster in the state in nearly a century (last being in 1924), one-sixth of the total population was directly affected by the disaster.More than 350 people have died and more than a million have been evacuated and the flood destroyed 906,400 hectares worth of crops. According to India Meteorological Department, Kerala received 2346.6 mm of rainfall from 1 June 2018 to 19 August 2018 in contrast to an expected 1649.5 mm during this season.The rainfall over Kerala during June, July and 1st to 19th of August was reported to be 15%, 18% and 164% respectively, above the normal. The first phase of flooding occurred towards the end of July. Subsequently another spell of intense rainfall was recorded at several locations on the 8th and 9th of August 2018. According to the report of Central Water Commission, places like Nilambur in Malappuram district, Mananthavadi in Wayanad district, Peermade, Munnar and Myladumpara in Idukki district and Pallakad in Palakkad district respectively registered a rainfall of 398 mm, 305 mm, 255 mm, 254 mm, 211 mm on 9th August. During this late phase (August 8to10, 2018) of high rainfall, flood conditions escalated on to Mananthavadi and Vythiri in Wayanad district.

The mountainous parts of the state witnessed widespread landslides and slope failures resulting in nearly 500 deaths and damage to property that included collapsed bridges and inundated cities and villages. Even after the floods the distress seemed to be continuing in the form of an unusual lowering of water level in the rivers, and wells drying up and caving. Slow slips seemed to be continuing to happen in some hilly locations (mostly in the district of Idduki) where more than 1,000 landslides had occurred. The water levels in majority of dams were almost near their Full Reservoir Level (FRL) and 35 of Kerala's 54 dams that had been overflowing had to be opened all at once, exacerbating an already dangerous situation to the level of a great calamity. The atypical second spell of rains from August 10, which normally should mark a waning phase, is what made the 2018 monsoon atypical, which had worsened the flood situation.

Nature in its destructiveness did not discriminate between mansions, even if they were made of imported marble, and the hutments of the poor. In the exclamations that were heard on social media videos from those witnessing the slow slide of large houses down slopes, followed by complete silence as the structures disappeared beneath piles of debris, one could sense the end of a dream - the opulence of the Gulf-returned *nouveau riche* Malayalis. The flood also washed away all the hierarchical barriers of caste and religion, with the people forced to rediscover a sense of community, assistance and accommodation - together typifying the sort of social character the state lost sometime in the late 1960s. Has Kerala learned anything from this unprecedented disaster? The silver lining seems to be that the post-flood Kerala is witnessing a healthy debate on how to rebuild the State that too following a sustainable growth trajectory and maintaining an environmentally healthy landscape that allows for the unhindered flow of rivers.

In an unfortunate turn of events, the social cohesion and altruism shown by the Kerala society during the days of disaster proved to be only skin-deep, as some of the recent events seem to suggest. Kerala's reputation as being a forward looking and evolved society has indeed taken a beating and its society now appears highly polarized on the caste lines. Clearly this situation does not augur well for a society that has to get ready to meet immense challenges, both in the social and economic fronts, which are interwoven with the sustainability of the environment. In the wake of future extreme climatic events, it is imperative that all elements of the society including its social, cultural and political classes embrace changes beyond their differences in their respective dicta.The society cannot afford to be distracted by politically motivated non-issues that can only be deterrents in the path to progress. Kerala must return to its constructive path towards realizing such causes as it had done in the past moments of crisis.

In this essay we will briefly look at the factors of why an excessive spell of rainfall had evolved into a flood of disastrous proportions and discuss why we need to look at new measures to mitigate the impacts of any future hazards of this sort especially in the background of climate change. We will be looking at some of these alternate options in our quest for the reconstruction of a new Kerala that will also constitute a post-flood vision for new Kerala.

Climate Change: A New Normal

While the incidence of extreme rain events in India is on the rise, with further intensification expected by the end of the century, observations also indicate an overall decline in the mean rainfall levels. The weakening monsoon circulation, increasing frequency of El Niño events and accelerated air pollution and land-use changes are believed to be the major factors contributing to the decline in the rainfall. This juxta position of increased frequency of extreme rainfall events and overall decrease in mean rainfall suggests that alternating spells of flood and drought are going to be a fixture of the Indian subcontinent's future. Because of its geographical location and its densely populated coastal regions, Kerala must bear the full brunt of these changes as climate is likely to see-saw between these two extremes.

A normal monsoon in Kerala provides about 2000 mm of rain, but by middle of 2018 August, the state received about 2350 mm - an excess of 42 per cent. Thus, we cannot overlook the fact that the current flood was associated with an abnormal phase of monsoon. In a typical monsoon season, the state receives its maximum showers in the months of June and July. But this time, the showers not only spilled over to August, but it rained with an accelerated intensity during August 9 and 17. The rains received on these eight days that departed from the regular pattern by about 300 percent made all the difference.

The monsoon distribution, in general, is not spatially uniform in India and it shows spatial variation in its intensity. This variability of the monsoon makes the Indian subcontinent one of the most vulnerable regions in the world. It is the opinion of the experts that the monsoon variability has 'amplified' in the recent decades. Observational evidence indicates a decrease in the northern summer (June-September) mean rainfall.

Observations during the last couple of decades (1950-2015) suggest that there is significant decline of 10-20% over the central Indian region. The decline in mean rainfall is ascribed to a weakening monsoon circulation, which in turn is attributed to a "combination of factors" like the warming of the Indian Ocean, increasing frequency of El Niño events, increased air pollution and land use changes. The experts have also predicted an overall intensification of extreme precipitation over most parts of the subcontinent by the end of the century (e.g. Roxy et al., 2017, A threefold rise in widespread extreme rain events over central India, Nature Communications, vol. 18, and references therein).

There is also an increasing realization that these spells of extreme climatic events are linked to warming environment induced by human activity. A rapid warming trend is also associated with the equatorial Indian Ocean that contributes to the increase in moisture content in the atmosphere. Some studies have pointed to the role of the local surface warming over the Indian subcontinent that also increases the local humidity levels for the frequency of extreme events although the actual process is much more complicated with several feedback systems playing their individual roles. A major puzzle has been the contradictory situation in which the frequency of extreme events is sustained despite an observed decline in monsoon circulation. We must remember that 41 percent of India's districts were reeling under drought conditions when Kerala had a phenomenal downpour. The numbers of low-pressure systems that can evolve into depressions that influence the distribution of monsoon rains are below average this year. The researchers suspect that the increased warming up of northern Arabian Sea that strengthen the westerlies drive up surges of moisture supply leading to spatially variable extreme rainfall episodes in the Indian subcontinent.Given the uncertainties in climate science it is imperative that we need to make rapid strides in predictive models of these phenomena.Thus, describing the current Kerala flood as a "one-in-100-year flood event" may not be apt, as the global factors that affect the climate are changing and we are not looking at the same 100-year window in future.

Growing Environmental Vulnerabilities

The 2018 floods also educate us on how a climatic *phenomenon can* spiral into a tremendous environmental *disaster.* Floods are caused by climatic changes, but their effects are amplified by human-induced factors, in particular the unscientific use of the land, which is of course a limited commodity in Kerala. Over the last several decades, mostly because of the increasing population density, environmental changes have been so pronounced in Kerala, that much of its landscape has been altered beyond recognition. Rivers and their constituent components such as catchments and watersheds have been increasingly impacted by dams built for irrigation and hydroelectric purposes. Traditional industrial units and their attendant investments have shown very little regard to ecological sensitivity and with no serious laws to prevent such activities they contribute to the pollution of water and air.

Disregard to well-studied long-term environment protection programmes have received very little attention in the unabated onslaught on Kerala's natural systems. Thus, the 2011 report by Madhav Gadgil, a renowned ecologist, and his team that investigated environmental conditions in the Western Ghats was almost ignored. Their report, based on detailed investigations had recommended that the entire region be declared "ecologically sensitive", and that mining, quarrying and construction of dams be banned to conserve biodiversity. By failing to accept and implement this report, the Government of Kerala missed a historical opportunity to show its commitment to the protection of environment, which would have earned it larger dividends and global recognition. Anyone who visits these mountainous regions of Kerala, regarded as the hallmark of its natural beauty would agree that the unchecked 'developmental' activities have altered their faces, affecting their aesthetic appeal and adding to the ecological stress. Construction activities gained acceleration in recent years and the unabated quarrying have destabilized the slopes. For anyone who visits the landslide affected areas post 2018 rains, it is not difficult to understand how the excessive quarrying and unplanned land use practices have made these slopes vulnerable to landslides.

The factors contributing to the rapid environmental degradation are social, cultural and economical and they make a good case study on how these factors are interrelated. Most important among the driving factors is the exponential growth of real estate. It not only forced the conventional agricultural sector to regress but it also led to anthropogenic activities that started to rapidly interfere with the ecosystems. The contribution of agriculture to the State's economy, which constituted about 60 percent at the time of formation of the Kerala is now reduced to a mere 10 per cent. Undoubtedly, the social and economic changes in Kerala that followed the Gulf boom were major catalysts in its transformation. The demand for increased construction led to more quarrying and mining of the river sand. Changing styles of livelihood, especially decreasing interest in agriculture led to alternate land use practices and there was rampant reclamation of wetlands and paddy for construction purposes. Eventually, Kerala, known for its varied ecosystems and biological diversity, gradually turned itself into a large, unitary metropolis, divorced from nature. If we scrutinize the State's built up area vis-a-vis its actual requirement, it is interesting that in a state with a population of three crore, there is a staggering sixteen lakh houses that are locked up with no one living there.

The most affected communities in this transition from a pro-agriculture to largely a service-centered commercial economy are the forest and coastal dwellers, whose livelihoods depend on the sustenance of their respective ecosystems.The forest dwellers (*adivasis*) are now being pushed to the edge, with inadequate resources to sustain their traditional farm based economy. As for the Government's schemes on welfare of *adivasis*, the history will record them as miserable failures. Together with the depletion in the quality of the environment there is also the threat from the spread of rare, communicable diseases recently reported in the state which is regarded as a consequence of its degrading natural systems.

Lessons Learnt from the Disaster and the Way Forward

So, what are the lessons we have learned and what is the way forward? It is important that for the hilly districts of Idukki and Wayanad, both the local and state authorities should rely on scientific reports on the landslide vulnerabilities to reach decisions on land allocations for various constructions. In these decisions, local soil properties and slope stability should be important factors to be considered in arriving at decisions, rather than political expediency.It is important to bring in restrictions on machine-mediated quarrying activities. A blue print that demarcates areas suitable for habituation and those to be left untouched should help strictly implementing the basic tenets of land zonation. A comprehensive masterplan on land utilization strategy based on a clear environmental vision needs to be prepared at macro and micro-levels to ensure that encroachment is minimal. These documents should contain clear guidelines for constructions, including recommended designs of houses that will match with local landscape and scenery.

The Government's lackadaisical approach is causing a slow death of yet another equally important ecosystem of Kuttanad in the coastal part of central Kerala - a region located 2 to 0.5 m below the sea level that suffered the most during the recent flood. The Kuttanad Package submitted by the M.S. Swaminathan Foundation was meant to be an action plan to reinvigorate paddy cultivation, along with the promotion of aquaculture, cattle and vegetable farming without generating much ecological distress. One of the important goals of the proposal was to reduce the impact of flooding. Like the Western Ghats, Kuttanad is also subjected to land grabbing (mostly by reclamation of estuary) and constructional activities by the real estate mafia in Kerala. According to a report prepared by the Sastra Sahitya Parishad, the estuarine part of the area has been reduced by 40 percent and the depth is reduced by 3.5 m. All these changes are affecting the movement of water and contributing to the flood havoc, not to speak of the alarming deterioration of biological diversity that will ultimately affect the livelihood of local population.

Talking of the movement of water, it is high time that we re-evaluated the role of Thaneermukkom bund in regulating the flow. It is unclear how a structure originally meant to be a 'regulator' has tuned to a 'bund'. It is pointed out that in the first 14 years the 'bund' was opened only for 14 days and during the six years that followed, it was kept open for 44 days. With the result, the estuarine part on the southern part of the bund has lost all its salinity and it has turned into a fresh water pond that facilitated the growth of invasive bugs and vegetation. The hindrance caused by the bund has also made the estuarine part a cesspool of toxic waste of fertilizers and pesticides. Tourism centered on houseboats has also contributed to the decline in ecological sustainability. There are about 1000 houseboats that operate in the backwater, in a space that can support only about 100 houseboats.

In the post-flood context, we also hear a lot about rebuilding Kerala, but there seems to be a lack of direction on what sort of developmental model the state should aspire to, particularly against the backdrop of global climatic and ecological changes. Although Kerala on its own cannot be a major player in the mitigation of climate change, it can do a lot to reduce its impact locally. The state must think seriously about how its water bodies can be saved from further encroachment and pollution. A way of turning the wetlands as a great resource is to use them as carbon storage centres and help reduce the amount of carbon dioxide in the atmosphere. Clearly these are innovative steps towards reducing the carbon footprint, an important agenda in the action plan for tackling the global climate.

Scientific conservation methods to protect the forests and biodiversity in the Western Ghats, which is subjected to encroachments mostly mediated through political networks, should be an important consideration.

That a lot of the recent controversies and scandals in the state have been centered on the legalities of settlements in hilly regions itself belies the seriousness of this problem. The political parties of Kerala have let themselves be manipulated by various business interests in rejecting the Gadgil Committee's recommendations, which provided a clear and scientific roadmap to preserve the stability of, and promote ecosystem services in the Western Ghats. Furthermore, it would have given the local bodies and the Panchayats enough power to decide on such issues and that would have also curbed corruption and nepotism to a great extent.

As such floods could recur, the state will have to develop more efficient warning and alert systems. Surprisingly, despite having a set of government-funded scientific institutes, the state did not have proper flood-maps that could have helped send more accurate alerts to the people. It is a matter of concern that very little scientific data is being collected on various phenomena that occurred *during* and right after the flood. It is important that the government agencies generate flood maps for all the river basins in the state and keep updating them. More effective communication methodology needs to be developed to improve flood risk literacy. Regional flood maps serve as a great medium for public engagement for communicating the magnitude of the risk and to generate cautionary warnings.

Finally, it is also a fact that the dams in Kerala, like elsewhere in India, do not have *rule curves* - the operating criteria, guidelines and specifications that govern the storage and release functions of a reservoir and which would be particularly useful during flood conditions. Although the Central Water Commission has a different view many experts have suggested that the deluge is not entirely of nature's making, and the impact is amplified because of faulty dam management. Instead of releasing water slowly at regular intervals dozens of dams (35 out of its 39 dams) were opened at one go when water reached danger levels, which led the flood waters to reach the nearby regions at a much faster rate than expected.It is high time we developed scientific management protocols for reservoirs. We need to factor-in meteorological forecasts into the computer models to predict the rate at which water flows into reservoirs and how much water needs to be stored. A table in the performance audit report (CAG) shows that out of 61 dams in Kerala, none had Emergency Action Plan or O&M Manuals. Kerala had also stated, in response to audit query from CAG of India, "No dam-break analysis was conducted in respect to any of the 61 dams in the state."

In the backdrop of human induced changes, it is more likely that we will be witnessing such extreme events more frequently. As geologists like Prof. Daniel Parsons (University of Hull, UK) has pointed out, "One of the most striking things from the videos and images emerging from the area is the brown colour of the flood waters and the extreme damage caused by landslides.... that geomorphology - the processes of erosion and deposition that shape the Earth's surface - is sensitive to rainfall intensity, so more frequent and more extreme floods mean more rapid changes across our landscapes". He further says, "Flooding is a challenge across individual, local, regional and global scales, and is set to increase in the future and its impacts will become more damaging. We need solutions across each of these scales to improve individual and societal resilience - so when flooding does occur it isn't the disaster we are currently witnessing unfold in Kerala". When the flood occurred, a general tendency was to describe it as "the worst in 100 years". This may not be a suitable descriptor of flood risk as it fails to communicate the newly discovered uncertainties inherent in the changing climate and the changing patterns of rainfall. We no longer can assume a "static, steady-state system" associated with the response pattern of the rivers.

A Green Future

Kerala is endowed with variety of natural landscapes and our natural environment is our most precious inheritance. We have a commitment to safeguard them for the benefit of future generations. Much of the success in this regard will depend on how to maintain cleaner air and water along with a thriving greener country that will be a safe haven for animals and plants. In the process we will have to learn how to use our land sustainably. Why don't we use the post-flood period as an opportunity to evolve an Environmental Plan for the next 25 years that will enunciate our long-term approach to various issues including land utilization, soil degradation, recycling, resource utilization, minimizing waste, extreme weather conditions (flooding and drought)? Taking a leaf out of United Kingdom's 25-year environment plan, such a document can come up with bold initiatives on environmental land management, new utilization rules for water, reducing the environmental impact of pesticides, introducing technologically driven new agricultural practices, introducing methods for soil health, restoration of water bodies and wetlands, developing flood management solutions, maintaining the sustainability of drainage basins, increasing and incentivizing water efficiency, development of geo-heritage sites and parks, helping schools to generate nature-friendly campuses, achieving zero-avoidance of plastic waste, recycling and usage of waste water etc. This plan can be developed as an extension of the existing "Green Kerala Mission" that is already being drawn up by the State Government.

Overall, and all things considered, the disaster should be taken as a grim reminder that any developmental models the state adopts must take the vagaries of the climate, which have become the new norm, into account, as well as strive to make human progress without triggering ecological disaster. At a fundamental level, the lesson from the 2018 floods is clear in that any future developmental plan must take the uniqueness and diversity of the environment into account. Ultimately our success will be evaluated on how we manage our limited resources for the welfare of the people without sacrificing the environmental sustainability. This goal can be accomplished only by implementing an advanced environmental protection system through the development of non-polluting industries, adoption of alternate energy sources, waste recycling, sustainable agricultural practices and a strong ecotourism industry supported by innovative legislation and linking conservation and reforestation.

Such alternate "degrowth" options have to be considered more closely in the quest for the post-flood reconstruction of a new Kerala that has now provided an opportunity to set a new example for India. In the coming fifty years the world that presently remains beholden to fossil fuel mode of production will slowly emerge into ecological economics. And, Kerala with its collective wisdom and its unique social evolution should be in the forefront of that revolution and set an example for the whole country. The 2018 event is a grim reminder that the climate change related disasters are here to stay and that any future developmental models cannot ignore this aspect. It has become a new norm that must be taken into account for the sustainable progress of humans without triggering ecological disasters. The 2018 flood provides us an opportunity to right a wrong that we have committed in the past.

9

August 2018 Floods and Landslides in Kerala: From Emergency Response to Reconstruction and Recovery

N. Vinod Chandra Menon

Introduction

The monsoon entered Kerala on 29th May 2018, and Kerala had witnessed relatively severe pre-monsoon showers before the regular onset of the monsoon season. The Indian Meteorological Department (IMD) was monitoring the rainfall as usual and was issuing routine alerts and warning depending on the rainfall pattern in various parts of the country. The rainfall was high in June and July 2018 and in a few villages in Wayanad and Idukki, landslides also happened because of the high rainfall. The water levels in the dams in Kerala were reaching high flood levels in August and the Indian Meteorological Department (IMD) had issued alerts and early warnings indicating very high rainfall in the first and second weeks of August.

During the period 8th to 15th August, the normal rain in Kerala was expected to be 98.4 mm, but actual rain during 8th to 15th August 2018 was 349.7 mm, indicating an excess rain of 255%. On 16th August, the normal rain was expected to be 13.5 mm even though the actual rain on that single day was 137 mm, indicating an excess rain of 915%. On 17th August, the normal rain was expected to be 14.5 mm but the actual rainfall was 76 mm, indicating an excess rain of 424%.

In Independent India, no state has ever before witnessed all the districts in the state facing a Red Alert issued by the Indian Meteorological Department (IMD) as was seen at 1.30 pm on 15th August 2018. The following table indicates the alarming situation caused by the unprecedented heavy rainfall during the month of August 2018. The excess rain during these days created stress on the dams in Kerala which were already facing high Flood Risk Levels, forcing the authorities to take the decision to release the water from the dams.

Aug 8 - 15		Aug 16		Aug 17	
Excess Rain :	255%	Excess Rain :	915%	Excess Rain :	424%
Normal	Actual	Normal	Actual	Normal	Actual
98.4mm	349.7mm	13.5mm	137mm	14.5mm	76mm

Rainfall figures are statewide averages

The authorities took the decision to release the water from the dams on 15th August after the red alert was issued for all the 14 districts of Kerala by the Indian Meteorological Department (IMD) because of the very high rainfall threatening to overtop the dams where the water levels were already reaching high flood levels. All the dams in Kerala opened the shutters and a huge quantity of water was released from the dams breaching the river banks, destroying crop lands and damaging houses, roads and bridges and uprooting trees, lamp posts and power supply transmission networks which also resulted in disruption of water supply in several flood affected districts. In some parts, the water was released from the dams without providing advance warning to government officials, elected representatives and local level functionaries causing panic among communities threatened by the rising water levels in the rivers and even in the urban and peri-urban areas in the catchment areas of the dams.

The initial estimates of the areas and population affected suggested that about one-sixth of the population of Kerala must have been impacted by the floods and landslides. It was confirmed much later that 1259 villages out of 1664 villages (75.66%) in Kerala were impacted by the floods. Out of the total 941 Grama Panchayats in Kerala, 511 (in the districts of Kottayam, Pathanamthitta, Alapuzha, Ernakulam, Thrissur, Idukki and Wayanad) were the worst affected. Kerala has 941 Village Panchayats (Grama Panchayats), 152 Block Panchayats and 14 District Panchayats; in the urban areas it has 87 Municipalities and 6 Corporations, a total of 1200 Local Self Government Institutions. The number of towns in Kerala increased from 99 in 2001 Census to 461 in 2011. The number of Urban Areas in Kerala has now become 520. The increasing trends in urbanization has seen the simultaneous increase in the aspirations of people to build pucca houses with RCC roofs in the urban, peri-urban areas and even in the rural areas. The foreign remittances sent by the people of

Kerala from abroad, especially from the middle east countries, have further accelerated investments in pucca houses in most parts of Kerala.

There was overwhelming response in emergency search and rescue and evacuation by elected representatives, government officials, youth and first responder agencies including the armed forces personnel, NDRF personnel, coast guard, police, fire and emergency services, fishermen, Kudumbashree members, volunteers from Indian Red Cross Society, and general public from diverse sections of the society. Relief camps were established to provide shelter for more than 14.5 lakh people who faced the onslaught of the floods and landslides. In one of the largest rescue operations mounted in natural disasters in any state, the concerned Ministries of the Government of India deployed 40 helicopters, 31 aircrafts, 182 teams for rescue, 18 medical teams of defense forces, 58 teams of NDRF and 7 companies of Central Armed Police Forces along with over 500 boats and necessary rescue equipment.According to the state government's estimate, a total of 4,537 fishermen participated in the rescue operation with 669 fishing boats and managed to rescue more than 65,000 marooned people from various districts.

The Government of Kerala announced that Rs. 10,000 will be given as emergency cash support to the families of people who took shelter in the relief camps. The flood affected people were also provided Family Kits with food and Non food items and "Back to Home Kits" when they returned to their homes from the relief camps after the flood waters receded. The uninsured losses to agriculturists, artisans, petty traders, micro enterprises and small businesses appear to be enormous and these people will find it difficult to restore their livelihoods in the absence of appropriate support.

The Cochin International Airport Limited (CIAL) was inundated with the flood waters and the aircrafts were grounded. The airport was not able to operate for a few weeks. The Cochin airport is in a low lying area and the Disaster Management Plan for the Cochin International Airport (CIAL) and similar critical infrastructure must be reviewed and revised to ensure that appropriate Standard Operating Procedures are incorporated in the Disaster Management Plans of such establishments.

The high rainfall in Kerala from 8th August 2018 onwards leading to extremely high water levels in the dams in Kerala was described as the cause of the unprecedented floods which affected several million people across the state. The high water levels in the dams, combined with heavy inflows into the reservoirs and extremely heavy rainfall forced the opening of shutters in all the dams simultaneously in Kerala.

The weak areas of governance which resulted in the unprecedented flood disaster, reportedly the worst in more than a century, cannot be overlooked.Changes in land use, quarrying, sand mining, stone cutting, blasting, violation of construction regulations on the river banks, silt deposits in rivers and water bodies, deforestation, etc. increased the ecological fragility in the eco sensitive areas. The coordination between various agencies of the Government of India and agencies and institutions of the Government of Kerala must be strengthened to avoid the recurrence of such incidents. Early warning systems, accurate forecast using Doppler radar data, Automatic weather stations, micro weather tracking and nowcasting using remote sensing and satellite imagery and weather variability monitoring models need to be strengthened. The convergence of Climate Change Adaptation (CCA), Bio-diversity Management and Disaster Risk Reduction (DRR) Practices and mainstreaming them with the local development plans at the district, gramapanchayat and village levels must be ensured.

The release of enormous volumes of water from the dams caused land subsidence, fissures, decline of ground water aquifer levels, disappearance of water from wells, changed course of rivers, river erosion and damage to infrastructure, assets and amenities. The loss of lives, disruption of livelihoods, collapse of traditional cottage industries like handloom and handicrafts and the misery faced by the flood and landslide affected communities was unexpected and the people were not prepared to face such an unprecedented flood.

The Impact of the Floods and Landslides

It has been reported in the media that the floods in Kerala caused damage to Panchayat Roads to the extent of more than 82,000 kms while the roads maintained by the Public Works Department saw damage to over 14,000 kms of road length. More than 221 bridges were reportedly damaged by the floods. The crop damage in the agriculture sector was estimated to be over 52,000 hectares. The power supply, water supply and telecommunication infrastructure also suffered damage and destruction. The drinking water sources like wells were damaged due to contamination with flood waters and due to the overflowing of septic tanks.

The World Bank and the Asian Development Bank made a preliminary assessment of the damage caused by the floods at Rs. 25,046 crore. The World Bank and Asian Development Bank team covered the flood impact on 11 key sectors like housing and public buildings, urban infrastructure, rural infrastructure, irrigation and water resources, power, transport, health, livelihoods, natural environment & bio diversity and cultural heritage. The damage assessment by the World Bank ADB Team across the various sectors shows the extent of damages.

In the Transport sector it was Rs. 8557 crore (Rebuilding state roads: Rs 7647 crore and Rebuilding National Highways: Rs 910 crore); In the Agriculture sector (Rs. 2093 crore); Livestock (Rs 154.7 crore); Fisheries (Rs. 225.4 crore); Industries (MSMEs) (Rs 641.9 crore); Handloom & Coir (Rs. 9.8 crore); Tourism (Rs. 676.9 crore); Housing (Rs. 2534 crore); Public Buildings (Rs 191.1 crore); Urban Infrastructure (Rs. 2093.1 crore); Rural Infrastructure (Rs. 5216.4 crore); Irrigation & Water Resources (Rs. 1484 crore); Power (Rs. 353.1 crore); Health (Rs. 280 crore); Livelihoods (Rs 3801.7 crore); Environment (Rs. 452.2 crore) and Cultural Heritage (Rs. 86.1 crore). The total damage estimated amounted to (Rs. 25,046 crore)

Special efforts must be made to seek the guidance of geologists, civil engineers, structural engineers, geo-technical engineers and hydrologists to assist the District Collectors in identifying villages and settlements which may have to be relocated based on the fissures, subsidence, debris flows, landslide history, drainage pattern and other relevant parameters. The Kerala floods and landslides have provided an opportunity for rebuilding Kerala.

A detailed Post Disaster Needs Assessment (PDNA) was carried out for the Kerala Floods, led by the Government of Kerala, with representatives of the United Nations, the World Bank, Asian Development Bank and the European Union Civil Protection and Humanitarian Aid. The PDNA assessed the damages and losses caused by the floods and estimated the total recovery needs as Rs. 31,000 crore. The break up of the PDNA assessment is given below.

Social Sectors

Housing, Land and Settlements: Rs. 5,443 crore; Health and Nutrition: Rs. 600 crore; Education and Child Protection: Rs. 214 crore; Cultural Heritage: Rs. 80 crore; Sub Total: Rs. 6,337 crore.

Productive Sectors

Agriculture, Fisheries and Livestock: Rs. 4,498 crore; Sub Total: Rs 4,498 crore.

Infrastructure Sectors

Water, Sanitation and Hygiene: Rs. 1,331 crore; Transportation: Rs 10,046 crore; Power: Rs 353 crore; Irrigation: Rs 1,483 crore; Other Infrastructure: Rs 2,446 crore; Sub-Total: Rs. 15,659 crore.

Cross Cutting Sectors

Environment: Rs 148 crores; Employment and Livelihoods: Rs 3,896 crore; Disaster Risk Reduction: Rs. 110 crore; Gender and Social Inclusion: Rs 35 crore; Local Governance: Rs 32 crore; Sub-Total: Rs. 4,221crore; Total: Rs 30,715 crore; Integrated Water Resources Management: Rs 24 crore; Grand Total: Rs 30739 crore, Rounded Off to Rs. 31,000 crores

Fund Mobilisation for Post Flood Rehabilitation, Reconstruction and Recovery

The resource requirement for recovery for addressing the damage caused by the floods and landslides was estimated at Rs. 31,000 crore in the PDNA by the United Nations agencies. The Prime Minister, Home Minister and MoS, Home in the Government of India declared allocation of Rs. 500 crore, Rs. 100 crores and Rs. 80 crore respectively after their visits to the flood affected areas. The Government of Kerala reportedly has Rs. 786 crore in the State Disaster Response Fund (SDRF) as allocation to the state from the Government of India. The funds expected from other sources are as follows:

It is expected that the Government of Kerala will be able to raise Rs. 7,200 crore from the World Bank, Rs. 400 crore from the Rural Infrastructure Development Fund, Rs. 2,500 crore from the National Bank for Agriculture and Rural Development (NABARD), Rs. 1,500 crore from the Panchayat Development Fund, Rs. 1,000 crore from the Goods and Service Tax Cess and Rs. 2,000 crore from the Plan Flood Deficit, making a total expected resource mobilisation of Rs. 15,900 crore. The Chief Minister's Distress Relief Fund (CMDRF) has also raised resources to the tune of about Rs. 3,400 crore. The post-flood rehabilitation, reconstruction and recovery is dependent on the availability of funds for undertaking the reconstruction of damaged infrastructure, destroyed buildings, roads and bridges and other amenities.

The Rebuild Kerala Initiative has set up a crowdfunding portal (see https://rebuild.kerala.gov.in) to mobilise funds for school rebuilding, livestock replenishment, house rebuilding, and for supporting anganwadis.

It is estimated that the Government of Kerala will collect about Rs. 15,900 crore through various funding options.

The Ministry of Rural Development, Government of India sanctioned Rs. 1800 crore under the Mahatma Gandhi National Rural Employment Guarantee Scheme (MGNREGS) for 2018-19 for Kerala for creating 5.5 crore person days of work.

Rich Social Capital in Kerala

About 76,285 houses were cleaned by 136,885 Kudumbashree volunteers. The Kudumbashree team cleaned 2,165 public buildings.1,773 families were provided with shelter by Kudumbashree members in their own homes. The Kudumbashree Constructions with all women team have built many houses for the families whose houses have been destroyed by the floods.

It has been reported that 4357 fishermen became miracle saviours of marooned people in the worst flood affected districts by bringing 669 boats on trucks from the coastal districts of Kerala and rescuing 65,000 marooned people and bringing them to the safety of relief camps. Elderly people, women, youth and people from all sections of the society irrespective of caste, class, gender and age came forward to extend help to the flood affected people. People from other states, Non Resident Indians from foreign countries and the Malayali diaspora abroad reached out by contributing in cash and kind to help the flood affected people of Kerala.

Community kitchens were organized by charitable trusts, Non Governmental Organisations and corporate sector entities. Hundreds of thousands of volunteers reached out in response to the call from young district collectors to extend their support to the relief efforts during the immediate post flood days. More than 80 humanitarian assistance organisations from different parts of the country came together and extended their support during the post flood relief and rehabilitation phase working closely with more than 182 local organisations in Kerala. Inter Agency Groups were established by Sphere India in 10 districts under the chairmanship of district administration officials starting from the first meeting of civil society organisations chaired by the District Collector, Wayanad on 16th August 2018.

The reconstruction challenges being faced by the flood affected communities are the following:-

The Land Use Plan Maps, Flood Inundation Maps and Landslide Maps in the worst affected districts were not available in most districts.

The study of changes in geo-morphology and hydro-morphology after the floods and landslides in the flood and landslide affected areas needs to be carried out to ensure that the reconstruction efforts are not rebuilding risk.

The Study of subsidence, soil piping, fissures, liquefaction, changing course of rivers and other trends in the flood and landslide affected areas are extremely important in the reconstruction and recovery efforts.

The Landslide Susceptibility and Slope stability in landslide affected areas also needs to be studied to ensure that the landslides do not recur in fragile areas during the coming monsoon months.

Relocation possibilities and cluster housing options must be explored in land scarce situations because the landslide affected areas have become extremely fragile.

In the flood and landslide affected areas, there is a need to shift from dependence on supply driven solutions to demand driven solutions through multi-stakeholder engagement and consultations, especially with disaster affected communities.

There is also a critical imperative to shift from reliance only on structural engineering solutions to increased acceptance of social engineering approaches through consensus building with disaster-affected communities.

A pragmatic reconstruction and recovery strategy must review land use maps, flood inundation maps, landslide maps and analyse the geo-morphology, hydro-morphology, satellite imagery and remote sensing, susceptibility and slope stability for considering the approvals for rebuilding of destroyed and damaged houses.

It has been found in the recent years that most of the funds earmarked for disaster management are being utilised by the state governments and Union Territory administrations for post-disaster relief efforts. There is a need for reviewing the allocation of funds for disaster management and strengthen disaster preparedness, disaster risk reduction and climate change adaptation in multi-hazard prone areas. It is seen that hydro-meteorological disasters are becoming more frequent and their intensity is increasing. Even though the lives lost in disasters is reducing over the years, the devastation caused by natural disasters on infrastructure, assets and livelihoods is increasing in an exponential manner.

The financial health of the Government of Kerala indicates that there is a need to evolve very radical and pragmatic steps for ensuring financial discipline in the state.

The Suggested Architecture for Nava Keralam

Communities must be centre-staged in community-led, owned and managed plan formulation, design, development, implementation and monitoring. Disaster Risk Reduction (DRR), Climate Change Adaptation (CCA) and Sustainable Development Goals (SDGs) must drive development planning at the local level.Special attention needs to be given to the marginalised and weaker sections, elderly, differently abled people, palliative care patients, etc.

Rainwater harvesting must be made mandatory for all new houses constructed in Kerala.Solar panels for individual houses and micro grids at the local levels with the possibility to sell saved surplus energy to the Electricity Board must also be considered. Buildings of all new public institutions like hospitals, health centres, schools, anganwadis etc. must be made disabled friendly. In coastal areas, mangrove plantations must be promoted with support of local Panchayati Raj Institutions. Appropriate solid waste management solutions must be adopted by Urban Local Bodies and Panchayati Raj Institutions through a multi-stakeholder engagement and inclusive consultative process.

The Kerala society showed during the floods and landslides in 2018 that humanitarian considerations are supreme in the face of the devastation which did not polarise or divide the disaster affected communities on caste, class or gender differentials. The society as a whole made special efforts to address the needs and concerns of people among the vulnerable sections. The elderly people, widows, palliative care patients and people with disabilities were taken care of by the volunteers and rescue teams. The psycho social care and mental health care of traumatised people who lost their assets, crop lands and houses received the attention of specialised institutions and specially trained mental health professionals.

The Nava Keralam initiative must be grounded on eco-friendly sustainable development and organic agriculture with special consideration of endangered bio-diversity in the Western Ghats in Kerala. All new construction in Kerala must be undertaken with building permissions granted by local bodies only on compliance of the provisions of the National Building Code 2016, building byelaws along rivers and coastal areas and development control regulations. Scientific application of tools and techniques for monitoring reservoir levels, inflow into the reservoirs from rivers, water levels in rivers and rainfall forecasts and "nowcasts" must be encouraged. Effective mechanisms for improving coordination between Central Water Commission (CWC), Indian Meteorological Department (IMD), Kerala State Electricity Board (KSEB), Kerala State Disaster Management Authority (KSDMA), Kerala Dam Safety Authority and Kerala Coastal Zone Management Authority and other relevant Central and State government agencies must be explored.

Improved surveillance and monitoring using sensors, models and satellite imagery and remote sensing, and application of Scenario Analysis, Expert Systems and Artificial Intelligence must be ensured. Rebuilding Kerala must build on inherent strengths of Kerala's **"social capital":** Kudumbashree self help groups, Community Based Organisations, *tech savvy youth*, start up companies, senior citizens, elected representatives of ULBs, PRIs, Gram Panchayats, District Panchayats and Legislative Assembly, cadres of political parties, fishermen, ex-service men, media, multi-disciplinary professionals, faith based organisations, spiritual leaders and their disciples.

Summing Up

Kerala's development trajectory from the hitherto pursued lop-sided development strategies must change to eco-friendly sustainable development pathways. This must be the new mantra of Rebuild Kerala Initiative. Sub-district level disaster management plans must be formulated at all levels by undertaking risk assessment, risk remediation and risk reduction approaches and incorporating disaster risk reduction, climate change adaptation and management of extreme weather events in micro level development planning. Resource Centres must be established at every block and Gram Panchayat level in worst affected districts with backward and forward linkages as has been attempted now with the establishment of Housing Facilitation Centres.

There is an urgent need for establishing a 2018 Flood and Landslide Museum and Recovery Lab with digital documentation of land use maps, flood inundation maps, landslide maps and archiving documents, reports, good practice insights, anecdotal documentaries etc. on the floods and landslides. "Smriti Vans" must be established in memory of those who lost their lives in the floods and landslides so that the memory of the floods and landslides will live through the ecological restoration efforts nurtured by the local communities as a grim reminder of the potential devastation that can recur if we take nature for granted once again. The repeat of the 1924 floods took 94 years. We must do everything possible to make sure that such disasters do not recur again. This can be ensured only through collective consensus for a new eco-friendly development pathway which puts the larger public interest as its foundation.

References

1. Government of Kerala, United Nations, World Bank, Asian Development Bank, European Union Civil Protection and Humanitarian Aid; Kerala Post Disaster Needs Assessment, Floods and Landslides August 2018: Thiruvananthapuram, Kerala.
2. Kerala State Planning Board, *Economic Review 2018*, Thiruvananthapuram, Kerala, January 2019
3. K.P.Kannan (Ed) (2018): *Kerala in Transition, Essays on Economy, Polity and Society,* Laurie Baker Centre for Habitat Studies & Centre of Science and Technology for Rural Development, Thiruvananthapuram.

10

Exacerbated Flood Damage in Kuttanad in 2018: Causes, Consequences and Caution for the Future

N.C. Narayanan

1. Introduction

Kerala was devastated by the floods of 2018 where most of the districts were partially affected. The unusually heavy rain induced by the low-pressure systems in the Bay of Bengal (which by itself is linked to warming of the oceans), the landslides that occurred in multiple areas along the Western Ghats and simultaneous opening of dams across the state are believed to have led to this unprecedented disaster. The antecedent rainfall received till July had given bountiful inflows into the major dams and at the same time saturated the soil limiting further infiltration. In the worst affected flood, the state has seen since 1924, more than 480 people lost their lives with a preliminarily estimated economic loss of more than Rs. 200,000 million. Out of the fourteen districts in Kerala 12 districts suffered major losses and one of the worst hit regions was the Kuttanad region. The floods saw a mass evacuation of about 2.7 lakh people from the Kuttanad region within two days[1] to 673 relief camps. Without getting into the current debates on relief and rehabilitation, we explore the historical roots of the increasing floods and place it within the larger political economy of development of Kuttanad. This is done with a review of the available literature on reclamations and other interventions in the region over a period of time. We focus on some of the writings two to three decades ago to highlight the academic insights ignored that could have pointed out better sensible planning to face future disasters. Section 2 explores the political economy of reclamations, especially the huge loss of water spread areas due to measures for facilitating double cropping. Section 3 illustrates the environmental consequences of rice-centric development of Kuttanad. Section 4 opens up the discussions to the larger physiography, land use interventions and the intricate relationship of the process and concludes with some concrete suggestions on measures as pointers to deal with disasters in future.

2. Development of a Rice-Centric Economy that Aggravated Floods: Political Economy of Reclamation and Shrinkage of Lake Commons

Kuttanad is a low-lying and water-logged region, transformed into a vast sheet of water of varying depths during the monsoon season (Pillai and Panicker, 1965:26-7). It is highly fertile land, replenished by silt brought by four rivers that originate from the western ghats; the area was found to be well suited to rice cultivation from early days.It is considered the only place in the world where rice cultivation is done up to 2 meters below sea level and is declared as a Globally Important Agricultural Heritage System by the Food and Agriculture Organization. Nearly 57% of Kuttanad falls under the Alappuzha district, 30 % in Kottayam district and remaining 13% in the Pathanamthitta district (MSSRF 2007). The region has high population density and rapid pace of urbanization with thirty seven out of the fortyfive Panchayaths (including municipalities

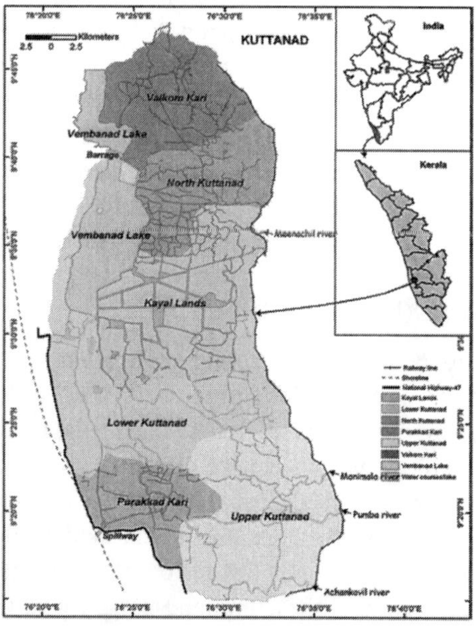

Figure 1: Kuttanad Wetlands [Source: Narayanan et.al (2011)]

and Kochi Corporation) having population densities of over 1000 persons per km2 with some exceeding 3000 persons/ km2 (Census, 2011). Three topographic areas identifiable in Kuttanad are the dry lands, the wetlands, and the water spread.[2] The wetlands are called *Padasekharams.* The rice fields reclaimed from marshes as well as backwaters are called *polders*. A sheet of water covers the agricultural land of Kuttanad when there is no cultivation. Outer bunds have to be built and the water bailed out in the beginning of every cultivation season. Reclamation of land for cultivation and flood control used to be undertaken by private farmers, with assistance from the state (Pillai and Panicker, 1965). Over a period of time Kuttanad became the rice bowl of the state and a rice-centric economy evolved here.

Lake Vembanad serves as an extensive nursery for marine prawns and also sustains a lucrative fishery. For centuries fishing has been an important occupation for the population of the area. There are both subsistence and commercial fisheries in the backwater area (KWBS, 1989; Kannan, 1979). Historically Kuttanad is a densely populated area by reason of the scope for multifarious economic activities like rice cultivation, fishing, coir making etc. The traditional agrarian structure until the mid-20th century was hierarchical and caste based; land being owned or possessed only by upper caste Hindus or Syrian Christians (Kannan, 1979). Records of reclamation are available from 1834 when the government of Travancore advanced loans for reclamation and by the dawn of the 20th century almost 2300 ha of Vembanad lake was reclaimed (GOK, 1971:5). The initiative for institutional reforms came from the monarchy and by 1850, as a consequence of this policy, a major share of the cultivated land and the whole of the wasteland came under the State. This led to the emergence of a class of independent farmers who reclaimed the backwaters of Kuttanad for rice cultivation through operations which required large capital investments (Pillai and Panicker, 1965).

We critically examine a pioneering study regarding this (Pillai and Panicker, 1965) to understand the process. According to the study, the increasing pressure of population on land during the last century and the exhaustion of shallow backwaters for reclamation purposes compelled the people to venture into the deeper waters of the Vembanad lake. These are known as the 'New Reclamations' (Pillai and Panicker, 1965:13). Without a discussion of the class dimension of the people who ventured in to this, it will seem that these are land-hungry peasants in search of extending the margins of agriculture for subsistence farming. The study placed the process in a global dimension and commented that "unlike the

European and Japanese reclamations, undertaken with the aid of the most modern techniques and backed by all the financial and organisational resources of the State, the reclamation of the Vembanad lake in Kuttanad was a masterpiece of private entrepreneurship undertaken by adventurous farmers with slender resources and crude techniques[3] (Pillai and Panicker, 1965:13).

Let us examine the story of the first reclamation in Kuttanad to understand the social relations of its origin and the active role of the state.

> The pioneer in Vembanad reclamation was one Eravi Kesava Panikkar belonging to the Chalayil family, one of the *leading landed aristocrats* in Kuttanad. He was 'an unusual person who *commanded great respect and influence.*' His very first effort bears an eloquent testimony to his innovative genius. It was a project to reclaim that portion of Vembanad which was situated at the mouth of the Cennankari river.... He proceeded to erect a barrier across the river at its mouth with local materials and manual labour.... *He had to face the opposition of the local population who filed a case against him for diverting the course of the river.* Sir Raja Rama Rao, the then *Dewan Peishkar, conducted an enquiry and personally inspected the site. But the farsighted Peishkar was more impressed with the potentialities of development, which this flash of genius had opened up than outraged by the offence of law. Fully alive to the problem of land shortage and food scarcity in the State, he saw in this novel adventure a new vista of possibilities.* Contrary to all expectations, instead of reprimanding the offender, the Peishkar congratulated Panikkar and *blessed his adventure*... Encouraged by the success of his first venture, Panikkar set about other reclamation schemes also. The legal proceedings against him proved to be a blessing in disguise, for he found in the person of his *defence counsel* Kavalam Neelakanta Pillai a *worthy partner* in this enterprise. The next block which he reclaimed was called after the statesman Peishkar Raja Ramapuram.... This is the beginning of the history of kayal reclamation which is not only highly romantic but from the point of view of economic development one of the most important things in Travancore History (Pillai and Panicker, 1965:6-17, italics added).

This story, apart from the romantics that the authors see, reveals some political economy dimensions of the early reclamations. The 'influential leading aristocrat' diverts the course of a river on which 'the local population' depends for their needs. The Dewan (Prime Minister of the then princely state of Travancore) makes personal inspection, and then congratulates him, leading to more reclamation with another 'influential' person who is a lawyer too. Naming the project after the Dewan was a way of showing gratitude. There was a repetition in the early 1940s when Thomas Murikkan was given permission to reclaim 1,800 acres of backwaters into three polders (GOK, 1971:6-7; Abraham, 1980:27). He named these polders after the Queen Regentess and her two sons. This is one of the reclamations that was attempted in the deeper parts of the Lake and that needed meticulous organization of resources including labour. There is no doubt about the adventurous nature of farmers' economic rationale behind this enterprise. The reclamation of land from water turned out to be more economic than the purchase of cultivable rice fields at high prices. The cost of reclaiming land in those days of cheap labour was much lower than the going price of rice fields and the entire initial investment on a reclamation project could be reimbursed from the net income over one or two seasons (Pillai and Panicker, 1965:20). The authors portray the initiative of the farmers as follows:

The initiative came from a few individual entrepreneurs with slender resources. The only aid at their command to conquer the invinsible Vembanad consisted of *crude indigenous implements and flimsy materials.* It is a miraculous feat that the*y* conducted the exacting operations of bunding, draining, ploughing and irrigating this lake area with the help of country canoes, wooden wheels, ancient ploughs and emaciated bullocks (Pillai and Panicker, 1965:21, italics added).

However, we note the huge lapse of not even mentioning the role of labour in this statement and the previous one. Who are 'they' who conducted the operations? The technology was crude, mainly a chain of water wheels used to drain the water with labour playing the most crucial role.[4] The processes by which the impoverished labour was drawn into the adventures of the wealthy farmers are yet to be written in the studies on Kuttanad history.

The role of the state according to Pillai and Panicker was 'shy and passive at the beginning'. Later, according to the study, the State also encouraged the movement by simplifying the procedure for *appropriating* backwaters for reclamation purposes…. Though the State came into the picture at an early stage, the role of the State was passive and shy. In recent times the State has taken an active interest in promoting this enterprise. Among the encouragements given are the *exemption of reclaimed lands from taxes during the first five years after reclamation, concessional rates of taxes during subsequent years, loans for cultivation and subsidies for pumping* (Pillai and Panicker, 1965:20-2, italics added).

The ecological threat to the Cochin Port noted by the Madras government led to suspension of reclamation activities for a nine-year period from 1903. Later continuation, strongly encouraged by the Travancore government, is mentioned in all studies (TSM, 1940; GOK, 1971; Pillai and Panicker, 1965). This reveals the powerful interests involved in the enterprise, able to influence and change government policies. A government publication also proudly claimed that the reclamations were the result of a steady policy inaugurated by the government and that successive Dewans sought to continue with such conversions to rice fields (TSM, 1940). It is important to note that the government, by allowing this 'appropriation by concessions and loans' of land in fact provided the capital for the whole enterprise. 'Capital' however, as Marx argued, *is not a thing, but rather a definite social production relation*, belonging to a definite historical formation of society, which is manifested in a thing and lends this thing a specific social character. Capital is not the sum of the material and produced means of production. *Capital is rather the means of production transformed into capital*, which in themselves are no more capital than gold or silver in itself is money. It is *the means of production monopolised by a certain section of the society*, confronting living labour power as products and working conditions rendered independent of this very labour power, which are personified through this anti-thesis of capital (Marx, 1974:814-15, italics added).

Thus the history of reclamation of Kuttanad also provides the background of the powerful landed interests, landless workers and crude technology in which the rice economy evolved. The introduction of motorized pumps in 1912 was a major impetus to the reclamation process (GOK, 1971: 8). A British engineer who visited Kochi saw the plight of Kuttanad cultivators, brought some kerosene engines from England and modified these to a 'box and para' pumping facility. The model is even now prevalent with later changes to crude oil, diesel and to electric motors from 1950 (Abraham, 1980:24). However, other technological developments were minimal in Kuttanad up to the introduction of Green Revolution technologies in the 1960s. The initiation of mechanical pumping and the advent of electricity to aid the farmer extended the scale of operations. From a farmer perspective it continued to be a risky operation and from 1910-1940 many such entrepreneurs incurred heavy debts and even went bankrupt, until the government of Travancore brought in the Debt Relief Act and saved many of them from complete destruction (Abraham, 1980: 21).

The cultivation was done only in alternate years with a fallow time of 10 to 20 months, but from 1940, with the flow of rice from Burma disrupted, the state banned fallowing and insisted by law on annual cultivation (Abraham, 1980:26). The acute scarcity of food grains during the period of World War II and the 'Grow More Food' campaign gave impetus to reclamation and by 1945, 8,000 ha of Lake Vembanad were converted to rice fields (Pillai and Panicker, 1965:19). As encouragement to farmers, 25% of pumping costs were paid back to the farmers as subsidy (Abraham, 1980:26). Government began trying various measures to augment rice production by reclaiming new areas, including forested hills in uplands and backwaters in Kuttanad, the best example being the sanction given to the huge reclamation of 700 ha mentioned earlier. A measure that affected farmer profitability was the 'quota' that used to be collected from 1942-1953 from rice farmers, which was a specific quantity of rice at controlled price to be distributed through ration shops at reasonable prices to consumers (Abraham, 1980). However, the considerable rise in the price of rice in Kuttanad during the war period in the 1940s boosted Kuttanad cultivation; the farmers were able to compensate the losses in previous years, could mobilize a big marketable surplus and wage payments were converted to cash (Jose, 1979:4). Consequent to the reduction in area and depth, the total volume of the backwater lying between Alappuzha and Azhikkode has been reduced from 2.45 km^3 in the beginning of 20th century to 0.56 km^3 in 1985 (Gopalan, 2002), a decline of almost 77 percent. The estimates of sedimentation suggest a decline in the water holding capacity of the lake, which increases the frequency and severity of floods (KSSP, 2017).

Table 10.1 : Area reclaimed for various purposes from Vembanad Estuary

Period	Area Reclaimed (ha)	
	Agriculture, Shrimp farms, housing, etc.	Harbour and urban development
1883-1903	2226.72	
1912-1931	5253.15	
1920 - 1936		364.37
1941-1950	1325.00	

(Contd...)

Till 1970s	5100.00	
1970-1984	800.00	
1900-1984	1500.00	
1978		10.78
1981-1985		319.04
2000		25.00
Total	16204.87	719.19

Source: KSSP (2017).

To sum up the discussion, the land-use interventions were directly related to the reigning policies devised to address the developmental needs of different periods. The 'grow more' food campaign in the 1940s encouraged extension of cultivation to wastelands like forests and low-lying marshes, which was considered a sign of progress - of 'conquering nature'. The Green Revolution strategies, the building of engineering structures to boost rice cultivation were major interventions in the Kuttanad natural system. A major impact is loss of the backwater area due to land reclamation, which started over a century ago. The environmental consequences of these rice-boosting strategies became evident from the 1970s and a string of studies was done from 1978 onwards to study the various aspects of environmental degradation. The Vembanad Lake is part of the larger Vembanad-Kol wetland and the prime brackish water body of South India. It was acknowledged as a wetland of international importance under the Ramsar Convention in 2002 (the area is protected and consists of 1,591 km^2 - partly below sea level).[5] Shallow parts of the lake and its shores were converted into rice fields over the last century and consist of highly fertile parts of land (Narayanan, 2003). The Vembanad Lake was encroached on and its shores were privatised for cultivation in the last century. Major reclamation work was done in the early 1940s with the consent of the State during the 'grow more food' campaign after the food shortages during the Second World War (Pillai and Paniker, 1964). Thus, lake commons used for fishing were privatised and thus enclosed over a period. In Vembanad (as part of a pattern followed in many Indian lakes like Chilika in Orissa State), the strong governmental support and the shift in policies that consisted in granting private entitlements for agricultural or other activities led to the gradual enclosure of common lands that were previously used according to customary rights by the local population (Narayanan and Venot, 2009). The next section examines the environmental consequences of the development interventions. From the context of floods, the shrinkage in lake and wetlands surrounding it reduced the water spread area and cushioning effects during times of heavy deluge. This is one of the major reasons for the heightened damage in the region.

3. Environmental Consequences of Rice-centrism in Kuttanad

The second phase of development interventions in Kuttanad started in the 1950s with the Nehruvian state-designed and managed agricultural modernization. Kuttanad was one of the two regions in Kerala that witnessed the Green Revolution strategies through a policy of heavy subsidies to promote HYV seeds, which in turn necessitated the intensive use of fertilizers and pesticides. Rice cultivation, traditionally systematized in tune with nature's rhythms (of monsoon floods and summer salinity) was disrupted with the building of several engineering infrastructure, oriented to double cropping of rice. HYVs' low tolerance to pests and diseases and non-judicious plant protection measures have increased the incidence of pests (John *et al.*, 19903 Thomas *et al.*, 1993; George and Krishnakumariamma, 1993). Diseases and consequent crop loss have become quite common (Aravindakshan, 1990; IIRD, 1978). Crops which were effectively protected with four to five treatments per season had to be sprayed ten to fifteen times in subsequent years (Mohandas, 1983:1). However, a shift to this mode of cultivation led to the following consequences:

Hydraulic Control for Double Cropping of Rice

Almost all state-sponsored development interventions in Kuttanad which were oriented towards making it into a rice-centric economy exacted a toll on the ecosystem. A major environmental impact is the shrinkage of the backwater area due to land reclamation into polders for rice cultivation. Two engineering structures built to drain off the floodwaters and check saline intrusion were constructed for facilitating double-cropping as part of intensifying agriculture to double cropping, a part of green revolution. The spillway that started functioning in 1955 for the speedy drainage of floodwaters to the sea was found to be ineffective due to faulty design. It was supposed to drain 64,000 m^3/sec of floodwaters coming from the highlands through the four rivers to the sea, but less than 20,000 m^3/sec of water is passing through it (Kannan, 1979:421). Before the construction of the spillway, these waters used go to the Cochin estuary and flood the low-lying rice fields. But when it was confined to flow through the spillway, flooding now also began to occur in its immediate vicinity. Another suggested possibility is the rise in sea level due to the rush of water from the Indian Ocean to the Arabian Sea during the SW monsoon. Because of this, even if the spillway was opened, water did not flow towards the sea (IIRD, 1978:37). Abraham (1980:30) supported this and argued that the flood of 1964 occurred during the winter following the strong SW monsoon, drowning 40,000 acres of *punja* crop.[6] Another unintended consequence is the accretion of sand from the sea near the shutters of the spillway, which prevents water flow towards the sea (IIRD, 1978:38). This report mentioned the views with hindsight by the chief engineer about the technical and organizational lapses in construction of the spillway, lack of understanding of sea level changes and doubt about whether the location of the spillway was appropriate (IIRD, 1978:39). Abraham (1980:32) comments that the constructors were unaware of the basic causes of floods in Kuttanad and that the spillway is an 'eternal monument to the incompetence of administrators'. The lesson learned here is that insufficient knowledge, i.e. uncertainty about natural processes, led to faulty design and unintended consequences. This design fault in spillway has thus augmented the floods in Kuttanad that it was supposed to mitigate.

Salinity, considered as 'pollution'[7] by earlier planners (Pillai and Panicker, 1965) became the 'ecological master factor' through hindsight in later studies (IIRD, 1978:37; Abraham, 1980:30; KSSP, 1978:39; KWBS, 1989; Nair and Pillai, 1993). Staggering of the crop season and resultant lack of discipline in rice cultivation after the construction of the salinity barrier has affected the natural balance of the system (KWBS, 1989). The salinity barrier commissioned in 1972 to prevent salt water into the Kuttanad system during summer was supposed to facilitate double cropping of rice, but has led to the decline or disappearance of several fish species. The estuaries and the backwater systems of the Kerala coast are the nurseries of several species of marine shrimp. The construction of the regulator has severely reduced the backwater area available for the prawns to spend their larval and growing stages of life, and the prevention of the inflow of saline water into the lake during summer has led to the decline or disappearance of several fish species that grow in saline water (Jhingran 1975; IIRD, 1978; Kannan, 1979). Decline in the catch of fish has cut into the livelihood opportunities of fishermen - a sizeable section of the population.[8]

Toddy tapping (country liquor) from coconut trees is another of the livelihood sources seriously affected by the salinity barrier (KSSP, 1978;

IIRD, 1978). According to a toddy-tapping labourer, the decrease of salinity due to the barrier is very detrimental to coconut cultivation and toddy production. He cites an instance when during the drought of 1983, the barrier was opened early and there was salinity ingress into Kuttanad and toddy production tripled that year. Coir processing was also seriously affected.[9] 95% of those involved in this activity are women (IIRD, 1978:37, Abraham, 1980, KSSP, 1978). The annual ingress of saline water to the upper reaches of Kuttanad during the summer months tended to check the growth of many of the water weeds. The complete prevention of saline water ingress accelerated the growth of weeds like salvinia, causing problems to rice cultivation and inland navigation and also adversely affecting the mangroves on the bank of the lake.[10] Stagnation of water due to closure of the regulator and the consequent decline in salinity prevent the settling and growth of lime shells (Kannan, 1979:428). The growth rate of the black clam, major resource for the lime shell fishery, is adversely affected by the rapid decline in salinity (KWBS, 1989; Aravindakshan, 1990:7). The faulty designs of the spillway and regulator are mostly due to the uncertainties and lack of information regarding the environmental processes. In retrospect, we see the limitations of development planning without due consideration of environmental and social aspects of land use. Unfortunately, the unintended consequences become clear only later through new information on the environment.

From the point of floods, both structures have huge impact: the spillway with its optimal functioning and salinity barrier blocking the flow of monsoon stormwaters.

Permanent embankments and roads

A major problem faced by the cultivators in raising the second crop of rice was the flooding of fields due to breaches in the temporary bunds. The state government drew up a scheme to construct permanent, but submersible bunds in 1974. In earlier times, the embankments used to be repaired just prior to the cultivation season; this recurring cost was feasible due to the large holdings and cheap availability of labour. With the turn in economics and social changes, this became difficult. The permanent embankments in polders (built mostly by state aid, implementing schemes such as the Kuttanad Development Scheme) are a boon to the farmers and for small farmers, cultivation would have been impossible without these measures. However, the environmental consequences of permanent bunds are: (i) preventing soil-enriching floodwater silt deposits; (ii) making it impossible to flush out acidity; and (iii) preventing fish ingress and growth during the fallow period.

The last two decades saw phenomenal rise of roads in Kerala. This can be directly related to the decentralized governance through the three tier Panchayat Raj Institutions. The level planning that started with IX Five Year Plan from 1996 facilitated decision-making at the local level. The most visible intervention like roads that can appeal to a large number of beneficiaries becomes the development imagination of every elected representative. Kuttanad also was no different. A place where traditional mobility was by canoes and boats aspired for speedier road transport. The low allocations to roads ensured that required culverts needed for efficient drainages were not given priority. The combined effect of permanent embankments and roads led to serious fragmentation of the wetland ecosystem that affected smooth drainage, especially during peak monsoons when Kuttanad receives flood waters from the four rivers draining from the Western Ghats. Reclamation reducing water spread area and the serious fragmentation accentuated the effects of floods. With unprecedented rains, there was no wonder that August 2018 floods became a disaster.

This section has clarified the fragile balances of the natural system and how the hazards of short-sighted interventions (like boosting rice cultivation), have affected the livelihoods of many.

Tourism and its consequences

There has been a spectacular rise of the tourism industry in Vembanad Lake, an official site of the International Ramsar Convention on Wetlands in Kerala, in the past twenty years. Houseboat tourism emerged as the backbone of the tourism industry of the state in addition to the many resorts, which have sprung up around the shores of the lake due to the scenic beauty of Vembanad Lake. On the one hand tourism and the rise of the houseboat industry, with more than 1000 boats[11], has led to job creation, it also puts pressure on the environment even though the natural beauty was the original reason for attracting such investments. The major sources of pollution are solid and liquid waste discharges from the houseboats. The water front area of backwaters along the lake is one of the most sought-after area for real estate development for resorts and more elite housing schemes. Encroachment into the backwater started in the 1980s, which accelerated during the last 10 years. A vigilance enquiry is in progress as regards several apartments many of which have been constructed illegally including violating the floor space index/floor area ratio stipulated in the CRZ suggesting the powerful nexus of builders, officials and political leadership who have little concern for environmental protection (KSSP, 2017). It has reduced the lake area and severely altered canal courses in several parts of the region contributing to the flood damage.

4. Floods 2018: Needs to Understand the Bigger Picture

Blessed with 44 rivers, backwaters and inland waterways, 8.7% of the total geographical area of Kerala is considered as flood prone by the National Disaster Management Authority (NDMA 2008). This year, the India Meteorological Department (IMD) had issued an extremely heavy rain (over 210 mm) warning to the state ahead of the monsoon onset. The state disaster management authority secretary even observed that during the last eight years IMD's highest order of extremely heavy rain alert in Kerala ahead of the monsoon onset is rare[12]. Expert opinions suggest multiple factors for the 2018 Kerala floods. News reports as suggested by the Central Water Commission proposed underperformance of water management systems that spills the excess water from Vembanad Lake into the sea. Also, the inadequate and the low carrying capacity of the lake due to siltation might have worsened the impact of deluge[13]. On June 22, 2018 newspapers reported that Kuttanad have been inundated and the Revenue Department of GoK has opened 19 gruel centres in the region following severe flooding[14] and by July 18, the number of gruel centres opened in Kuttanad taluk stood at 375. In the week starting 12th July, rainfall in Kerala was 122% above the long period average[15], and Kuttanad reported the worst flood in the last two decades[16] with a mass evacuation of about 2.7 lakh people from the region[17]. About 673 relief camps were functioning in Alappuzha district alone.To understand the scale of devastation we have to grapple with a more regional and temporal analysis.Narayanan (1994) has detailed the links between physiography, land use changes and environmental consequences in Kerala 25 years back and the cumulative effects of this process illustrates the root causes of the flood damages:

Larger Land Use Changes in the Intricate Physiography of Kerala

Kerala is an elongated land mass having the Western Ghats along its eastern margin with a maximum elevation of 2670 m (Anamudi) and the coastal plains along the west bound in by the Arabian sea. The physiographic diversity, along with the climate and soil have endowed the state with a very rich flora and fauna. The state can broadly be divided longitudinally into three physiographic sub divisions, the highland, midland and the low-land.[18] The highland is generally thickly forested in its upper reaches, while in the lower ranges the forests are interspersed with plantations. On the western fringe of the State are the lowland and the coastal zone containing a string of estuaries, and back-waters of which Kuttanad is part of. The rolling hills of the midland lie between these. Due to its location on the windward side of Western Ghats, Kerala receives an average rainfall of 2800 mm. The diversity in the original rocks, variations of micro-climatic situations and differences in the intensity and type of weathering under variable terrain conditions have led to the generation of different types of soil in Kerala (Government of Kerala, 1982). Because of the undulating terrain conditions in midland and in parts of the lowland, and of the highly dissected nature of the highland (which also is due to the high surface slopes and rainfall), the soils of Kerala are susceptible to erosion and in extreme cases landslides under conditions of improper land use and inadequate soil conservation measures.[19]

The Western Ghats form the catchment area of all the 41 rivers in the State. Invariably all the rivers are dependent on the forests for a sustained flow. A vegetation cover presents an entirely different surface to solar radiation as compared to the barren soil. Further, the role of vegetation in reducing surface run-off is of particular importance in a region like Kerala, where the topography is rugged and the rainfall is heavy and concentrated during a few months. Existence of forests helps in reducing the peak flow and prolongs the duration of the flow (IUCN, 1980). Most of the flash floods arise from the rapid run-off from catchments where the vegetation has been removed. By regulating the peak flow, the forests check flash floods in the valleys during rains and prevents extreme drought conditions in summer. Further, by regulating the flow it helps to prevent the saline water intrusion in the lower reaches in summer, by maintaining a minimum water level (Government of Kerala, 1982).

According to Menon (1983), deforestation in the State had acquired a different dimension in the sixties and seventies which had largely been cleared for "developmental activities". But that itself had acted as an incentive for more and more forests to be cleared for agricultural purposes. Kannan and Pushpangadan (1988) have summarised the reasons for deforestation as; (1)encroachment of forests by powerful rural interests, making use of a large army of land-poor and land hungry peasants; (2)clear felling by government for raising plantations and construction of irrigation and hydro-electric projects. (3)illegal felling of trees and plundering of forest resources by private interests, mainly timber contractors.

Deforestation and raising of tubers like cassava and monoculture plantations accelerate the process of soil erosion and increase the possibility of landslides in the highland. The soil erosion triggered by the cultivation leads to an increase in the amount of silt carried by the rivers bringing down the storage capacities of the reservoirs (for irrigation and power generation), thus wasting public investment (Chattopadhyay, 1984). This reduces the sediment load in the rivers downstream, but in the midland and lowland portions, heavy sand mining is done which again reduces the sediment capacity. The depth-width ratio of the stream is also distorted during the process. Rivers retain their natural shape by carving off the banks along with the buildings and cultivated land. A study in Neyyar river, showed that the extent of sand removal was dis-proportionately higher than the feeble countering mechanism of sediment deposition by the stream. The deepening of the channel further resulted in the depletion of ground water from either side of the river (Thrivikramaji, 1986). In the lowland, wetlands are drained, cleared and filled for agriculture, human settlement and industrial purposes. Such acts are likely to disturb the ecological balance of the coastal zone. The unprecedented scarcity of water in the coastal zone, drying up of wells which used to be perennial are indications of this. Intrusion of salinity in the wells can be attributed to the unscientific reclamation of extensive areas of wetland and the indiscriminate deforestation along the Western Ghats. The reduction/ changes in vegetation and increase in built-up area increases the chance of flash floods in the basin. This is exasperated in the months of June to November when almost 90 percent of the annual discharge from the rivers takes place. This enormous seasonal variation in flow is a major factor that impacts the entire wetland system, including the efforts to alter the flow pattern. With the increasing human-land ratio, consequent pressure on land and short-sighted land use practices mentioned above resulted in varying degrees of environmental degradation. This is important since these units are parts of a single fragile system where tampering with one part will automatically have its repercussions on the others. Kuttanad thus comes at the receiving end being in the downstream of four river basins and thus faces the cumulative effects of the damages.

Path Ahead

The above description highlights the need for a more integrated approach for managing the heterogenous land and water system in the State. This needs more in-depth land use analysis at local and regional levels. An immediate need is an attempt of zonation of existing disasters to understand the range of vulnerabilities spatially and socially. Second is the stringent measures of regulation before losing the collective memory of floods. Regarding Kuttanad, the most important element is the need for increasing the water spread area, which has limitations because of the committed land uses in the hands of private individuals. However, the remaining public lands/commons have to be protected with commitment. One of the important tasks is to understand the drainage network and ease out the congestions, especially vacate the encroachments. All roads and permanent embankments have to be provided with adequate measures of water flow by constructing structures like culverts. The completion of new shutters in the salinity barrier has to be complimented with a proper operation and management plan to facilitate flow of flood waters during monsoon. More investigations are needed to make the spill way operations better. Flood levels have been recorded by various initiatives and using this a proper Digital Elevation Model study has to be done to delineate and zone the flood levels. This will help prioritise the areas and habitations that needs immediate attention during an imminent flood. Similarly, socially vulnerable households like those with bed-ridden patients that need special care and help during relief could also be demarcated. Chances of increasing climatic events like surge in precipitation or rise in sea level will directly affect regions like Kuttanad first. Preparedness is most important. "Living with floods" must be the slogan with early warning systems, new houses to be built above the flood level with stilts and preparing the population for floods.

Notes

1. https://www.thenewsminute.com/article/kerala-floods-water-not-receding-kuttanad-cut-rest-state-87439
2. Pillai and Pani ~~k~~ckaer, 1965; KSSP, 1979; KWBS, 1989. 'Greater Kuttanad' is defined as the area extending over 874 km^2 with garden lands; just over 304 km^2 at an average elevation of 1m above Mean Sea Level (MSL) and about 500 km^2 of area at 0.6 to 2.2m below MSL (Chattopadhyay and Sidharthan 1985).
3. The very first experience of reclamation, which was the general pattern of the enterprise, is described as follows: 'Whole stems of coconut trees were piled into two rows on the bed of the stream. A broad corridor, which was formed in this manner, was cemented with garbage, river sand and clay and it emerged as

a bund. The ring bund rising from under the water was like a dream come true. It was indeed a remarkable engineering feat, though a risky and expensive operation' (Pillai and Panicker, 1965:13).

4. Water was drained by manually operated water wheel. An array of wheels of different sizes was used, arranged one below the next. Planted at the extremity on the outer ring bund was a wheel with 8 to 12 spokes to operate the installation (Pillai and Panicker, 1965:20; GOK, 1971:8).
5. Various rivers originating in the Western Ghats flow into the lake, which is connected to the Arabian Sea through an estuary: the lake is under tidal influence and subject to seasonal fluctuations in salinity levels (Narayanan, 2003).
6. Curiously, another technological solution of pumping water to the sea with 40,000 hp electric pumps was suggested by some quarters in both the above-mentioned studies. Abraham (1980:32) ridicules this: 'if this happens, it will be one of world's utmost tourist attractions as a symbol of technological foolishness.'.
7. The rationale of the salinity barrier was to boost rice production, and salinity was considered 'pollution' in those times. An earlier study noted that 'during the summer months the flow in the rivers which feed the lake progressively dwindles, while the *water in the lake area gets increasingly polluted by the invasion of salinity* from the sea. This produces a scarcity of fresh water for irrigation even in the existing fields..... It is expected that with the completion of the spillway and the salt-water barrier an open spread area of about 20,000 acres (more than 8,300 ha) will be available' (Pillai and Panikar, 1964:172-italics added). Later studies brought out the problems of checking salinity (KWBS, 1989; Nair and Pillai, 1993).
8. KWBS, 1989. - The few available data showed that estuarine penaeid prawn catches used to vary from 1,500 to 2,000 tonnes/year in the period 1970-1982, while at the time of the study, yields have declined to about 1,000 tonnes/year (KWBS, 1989).
9. While retting of raw husks takes three months in saline water, it takes 10-12 months in fresh water and hence retting has been seriously hampered by the operation of the regulator. The quality of fibre is reported to be inferior compared to that processed in earlier days (IIRD, 1978:46; KSSP, 1978:39, Kannan, 1979:428).
10. The mangrove habitat is a unique blend of land and aquatic eco-system. It has very good land conservation characteristics and prevents soil erosion ~~in~~ at the lake banks (Arvindakshan, 1990:7).
11. Mr Er. Vinothu, Port Officer in charge in Alleppey, *interview* in Alleppey on 23 February 2013.
12. https://timesofindia.indiatimes.com/city/thiruvananthapuram/high-alert-in-kerala-with-extremely-heavy-rains-on-may-28-ahead-of-south-west-monsoon/articleshow/64322715.cms
13. https://indianexpress.com/article/india/how-keralas-largest-lake-worsened-flood-water-panel-5345888/
14. https://www.thehindu.com/todays-paper/tp-national/tp-kerala/six-relief-camps-in-alappuzha/article24226099.ece

15. http://www.imd.gov.in/pages/press_release_view.php?ff=20180719_pr_293
16. https://www.thehindu.com/news/national/kerala/worst-floods-in-last-two-decades-ravage-kuttanad/article24454838.ece
17. https://www.thenewsminute.com/article/kerala-floods-water-not-receding-kuttanad-cut-rest-state-87439
18. The classification of area is according to elevation above mean sea level (MSL). Low lands are those below 7.5 m from MSL, midlands between 7.5 to 75 m and highlands above 75 m from MSL. Low land covers 10%, midland covers 41.76% and high land 48.14% of the total geographic area (Government of Kerala, 1989a).
19. It has been estimated that out of a total area of 38.85 lakh hectares, 19 lakh hectare area is affected by soil erosion problems.

References

1. Aravindakshan, M, (1990): *The Kuttanad Ecosystem, Five Decades of Rice Research in Kuttanad,* Kerala Agricultural University, Thrissur-680 654.
2. Chandy, J. (2013): "Economics and environmental dimensions of backwater resources in Kerala: A study with special reference to Kuttanad region", *PhD Thesis*. Mahatma Gandhi University.
3. Chattopadhyay, S, (1984): *Deforestation in Parts of Western Ghats region (Kerala) India,* Trivandrum: Centre for Earth Science Studies.
4. George Alexander and P. R. Krishnakumariamma, (1993): *Trends in the use of ecologically hazardous inputs for rice in wetland ecosystem: A 10 year case study,* Rice in Wet Land Ecosystem, 250-253. Thrissur: KAU.
5. IUCN, (1980) (ed) : *World Conservation Strategy: Living Resource Conservation for Sustainable Development,* IUCN–UNEP–WWF.
6. Jhingran, V.G., (1975): *Fish and Fisheries of India,* p. 255, Hindustan Publishing Corporation, New Delhi.
7. John, P.S, P.J. Ittyavirah, M.S. Nair and N.K. Sasidharan, (1993): "Effect of agronomic practices on rice yield and the incidence of major rice pests", *In Rice in. Wet Land Ecosystem,* 242-245. Thrissur: KAU.
8. Jose, A.V., (1977): "Origin of Trade Unionism Among Agricultural Labourers in Kerala", in *Social Scientist*, 5 (12), July:24-44.
9. Jose, A.V; (1979): "Trade Union Movement Among the Agricultural Workers in Kerala- The Case of Kuttanad Region," *Working Paper No. 93,* Trivandrum: Centre for Development studies.
10. Jose, R.V, (1999): "Inland Fishermen and Inland Fishing: A Study at Neelamperror Village (Alapuzha District)", *Discussion Paper No.6, KRPLLD,* Thiruvanathapuram: Centre for Development Studies.
11. Kannan, K.P., (1979): "Ecological and Socio-economic Consequences of Water Control Projects in the Kuttanad region of Kerala", *Proceedings of the Indian Academy of Sciences*; Vol. C2, Part 4, December 1979: 417-433.
12. Kannan, K.P, (1986): "Of Rural Proletarian Struggles: The Mobilisation and Organisation of Rural Workers in Kerala, India", *PhD Thesis* submitted to the Institute of Social Studies, The Hague, The Netherlands.

13. Kannan, K.P and K. Pushpangadan, (1988): "Agricultural Stagnation and Economic Growth in Kerala: An Exploratory Analysis". *CDS Working Paper* No. 227, Trivandrum: Centre for Development Studies.

14. Marx Karl, (1974): *Capital, A Critique of Political Economy*, Vol.III, Moscow, Progress Publishers.

15. Menon, A.R.R (1983): "Tropical rain forest: Present state and future strategies", Vanarashmi: 4(10/11): 5-12. (*KFRI Scientific Paper* No. 120), Thrissur, Kerala Forest Research Institute.

16. Nair, P.V.R and Pillai, V.K, (1993): "Changing Ecology of the Vemnabanad Lake", *in Rice in Wet Land EcoSystem* (eds) R.R. Nair, K.P.V. Nair and C.A. Joseph, Thrissur-10, KAU. P. 280-285.

17. Nandakumar, T. (2014): *"Four districts categorised as climate change hotspots," The Hindu* [online] Available at: https://www.thehindu.com/news/national/kerala/four-districts-categorised-as-climate-change-hotspots/article5801125.ece [Accessed 7 Sep. 2018].

18. Narayanan NC (1994): "Sustainable Land use for Agriculture: A Local Level Approach with Reference to Paddy Cultivation in Kerala, *M Phil Thesis* submitted to the Jawaharlal University at the Centre for Development Studies, Trivandrum.

19. Narayanan NC (2003): *"Against the Grain: The Political Ecology of Land Use in a Kerala Region in India",* Masstricht, Shaker Publishers.

20. Narayanan NC and Jean-Philippe Venot (2009): "Drivers of change in fragile environments: Challenges to governance in Indian wetlands", *Natural Resources Forum*, 33 (2009), PP. 320–333.

21. Narayanan, S.P., Thomas, A.P. and Sreekumar, B., (2011): "Ornithofauna and its conservation in the Kuttanad wetlands: southern portion of Vembanad-KoleRamsar site, India", *Journal of Threatened Taxa*, *3*(4), pp.1663-1676.

22. Padmakumar, K.G., (2013): "Kuttanad-Global Agricultural Heritage: Promoting Uniqueness", Proceeding of Kerala Environ. Cong, pp.62-72.

23. Pillai V.R. and Panikar P.G.K. (1965): *Land Reclamation in Kerala,* Asia Publishing House, Bombay.

24. Thomas Jim, N. Mohandas and M.J. Thomas (1993): "Contamination of the irrigation water in rice fields due to insecticidal spray application", In *Rice in Wet Land Ecosystem*. 246-249. Thrissur: KAU.

25. Thrivikramaji, K.P, (1986): River Metamorphosis Due to Human Intervention in the Neyyar Basin, Kerala, Trivandrum: Dept. of Geology, University of Kerala.

26. Varma, Ajayakumar, (2016): *Eco-restoration of Vembanad Lake & connected river systems.*

Malayalam Publications.

27. Abraham, John, (1980): *Kuttanad*, Kerala Sastra Sahitya Parishad, Trivandrum.

28. IIRD, (1978): *Kuttanad: Prasnangalum Sadhytahakalum,* Kottayam: IIRD .

29. Kerala Sastra Sahitya Parishad (KSSP), (1978): *Kuttanad Padhana Report. Problems of Kuttanad - A study report*, Trivandrum: KSSP.

Government Publications

30. *Census of India* (Various Years).

31. Census of India, (2011): *District Census Handbook, Alappuzha. Directorate of census Operation*. Kerala. Available at http://censusindia.gov.in/2011census/dchb/3211_PART_B_ALAPPUZHA.pdf

32. GOK, (1982): *First Report on the State of Environment in Kerala,* State Committee on Science, Technology and Environment, Trivandrum.

33. GOK, (1971): *Report of the Kuttanad Enquiry Commission*, Government Press, Trivandrum.

34. Kerala State Disaster Management Authority (KSDMA), n.d. *Kerala State Disaster Management Plan Profile*.

35. National Disaster Management Authority (NDMA), (2010): *National Disaster Management Gudelines - Management of Floods.*

36. TSM, 1940, *The Travancore State Manual,* Government Press, Trivandrum.

Unpublished Literature

37. KSSP, (2017): *What future for Vembanad Ecosystem: From one crisis to another or crisis to sustainable management? Report of People's Commission on Vembanad Ecosystem.*

38. KWBS, (1989): *Kuttanad Water Balance Study-Draft Final Report*, BKH Consulting Engineers, Thiruvananthapuram.

39. Mohandas, N, (1983): "Strategy of Pest Control for Achieving Targets of Increased Rice Production and Productivity in Kerala". In the Proceedings of the *Seminar on Stagnation of Rice Production in Kerala.* Vellayani: KAU.

11

Understanding Disasters and Resilience: Towards Rebuilding Kerala

S. Janakarajan

Introduction

The problem of ecological and environmental degradation pose a severe threat to basic human living. Its ramification is indeed much more severe in the less developed countries that are afflicted with chronic problems of political instability, demographic pressure, lack of political will, high level of illiteracy, unceasing poverty, increasing degree of urbanization and urban poverty, rapid industrialization, lack of basic needs and basic infrastructure, women subordination, high degree of corruption at all levels, poor health care and poor social security system, high population density with poor rural and urban infrastructure and the looming climate change threat. The growing menace of environmental and ecological degradation needs to be addressed in this context.

Natural resources (or what may be called *natural capital*) are nature's gift to the humanity which is used in every economic activity. All critical functions that are vital for human existence such as hydrological cycle, decomposition and recycling process, crop pollination etc are the result of direct eco-system services which the humanity is enjoying thanklessly. The so called growth, development etc. has been made possible only because of services provided by different ecosystems. But do we ever respect our eco-systems? For most of us and in particular policy makers the natural resources are given and infinite. The state and central governments and those who drive policies from behind (including economists in and out of World Bank) are so carried away that the growth is fabulous and that poverty and malnutrition could be eradicated simply by pushing the growth factors upwards. But have they ever considered that the so called growth and development are only gross which did not include damage caused to our natural capital. Who will pay for it? The nature beyond a point will not recover and recycle. At that stage the ecosystems will misfire. How will the market take care of the system when nature hits back by way of "disasters"? So far, markets have failed to respond to disasters such as heat waves, droughts, floods and landslides.

Let's think of some key natural resources / endowment

- Forests and forest streams
- Small water bodies in particular tanks
- Rivers and river pollution
- Estuaries, brackish-water lakes and Creeks
- Mangrove forests
- Coastal ecology and marine ecosystem
- Groundwater depletion and contamination
- Soil moisture
- Soil pollution / erosion
- Wet lands
- Clean Air
- Carbon space

All these are eroding, shrinking and or lost permanently. The governments (both state and central) in India do invest a great deal every year towards rejuvenation and recovery of natural capital. Expenditure made towards recovery and rejuvenation are called investments and hence added into the calculation of GDP and growth rate of the economy, but unfortunately natural capital is never recovered but on the contrary they are eroding continuously.

The key question: What is the value of this natural capital which we are losing every year? How many trillion dollars? Even if you have money to spend, can you create them? These are baffling questions with no answers.

Ecological and environmental protection and security are as much important as that of economic development and growth. Every production (as well as consumption) involves use of natural capital. But the key issue is to reach a trade-off point between growth and ecological and environmental sustainability. This is the central issue which is also very challenging to achieve in a competitive world.

Land Use and Land Cover Change - The Key

That the changes in land-use over time can contribute significantly to regional climate change conditions are by and large unrecognized. Land - Water - human relations and interactions are the key for human civilization and socio-economic development. From a rudimentary human society which involved "hunting and gathering" to a most modern society in urban settlements, land use changes are the basis and played a key role. Today, urbanization, industrialization and all other needs such as development of rural and urban infrastructure demand land. And, the demand for land is rising progressively, more than ever before. Thus the land use change has become the central issue which is often unrecognized or taken for granted without realizing its socio-economic, ecological, environmental and climatological consequences. The key elements that are affected in the process of land use changes are declining area under forest, decline in overall bio-mas production, distortion of given natural drainage systems and watersheds, top soil erosion, changes in soil and soil moisture conditions, disturbance in the geological formations, disturbance in the wild life habitations etc. Most important however is the rise in temperature leading to overall climatological and rainfall conditions, changes in farming practices and agricultural employment, changes in conditions and characteristics of groundwater regime, livelihood insecurity etc. Therefore, it is the combination of all these factors which contribute to floods, landslides, and droughts.

In the arid and semi-arid regions (and conditions), the problem starts with rise in temperature, which may not necessarily be due to changes in local land-use and bio-mass conditions but predominantly due to land-use changes elsewhere in the region. In the particular context of South India it may be disturbance in the Western Ghats leading to temperature rise. But in the hilly and elevated regions, the issue starts with land-use changes leading to temperature rise, droughts and floods.

The key questions in this perspective are,

- *What are the distortions made to the Western Ghats in the last few decades?*
- *What are its impacts on surface run-off (high-current flash floods), landslides, bio-mas production, soil erosion, soil moisture conditions and agricultural production conditions?*

It may not be easy to answer these questions in a straightforward manner. But let me figure out the fundamental causal links that have contributed to devastating conditions in the State of Kerala. Yes, it is absolutely disastrous! The Kerala Chief Minister informed the State Legislature that the total loss of human lives was 483 and 14 people were still reported to be missing. The total economic loss was estimated at more than the Plan outlay of the state for the year 2017-18 which was Rs.26,500 crore. About 400,000 people were displaced due to floods and landslides. Thousands of cows and buffaloes, goats, sheep and poultry were also lost along with numerous pet animals. Indeed, the cumulative damage on the affected families due to human loss, psychological stress, health of the people, loss of properties, livelihoods and social and economic impact of displacement are incalculable.

Let us see in the next few pages the extent to which the conditions in the Western Ghats have been distorted and its consequences.

About Western Ghats

The Western Ghats, which is regarded as the Water Tower of South India, runs over a length of 1600 km, covering a total area of about 160,000 sq km. Western Ghats is one of the ecologically very sensitive zones in South Asia and one of the richest hot spots in the world for bio-diversity. This is also the place where all the crucial rivers of South India originate. Historical records show that about three quarters of the land area of Kerala was under dense forest up to 18th century (Kerala forests and wildlife department, 12 August 2015). Three of the world's *Ramsar Convention* listed *wetlands-Lake Sasthamkotta, Ashtamudi Lake* and the Vembanad-Kol wetlands-are in Kerala. Kerala's rainfall averages 2,923 mm (115 in) annually. Rainfall in some of Kerala's drier areas the average rainfall is only 1,250 mm (49 in); But the the mountains of the eastern Idukki district receive more than 5,000 mm of rainfall.

Primary Reason for the Disastrous Condition in Kerala in 2018

Declining area under forest and rapid changes in Land Use and Land Cover in Kerala

A study carried out by T V Ramachandra of Indian Institute of Science, Bengaluru, concludes that there has been a massive reduction in the area under forest since 1973. This study confirms that the area under forest in Kerala has shrunk to the extent of 9064 sq km (906,440 hactares) since 1973. According to a study carried out by the Centre for Ecological

Sciences, Indian Institute of Science (IISc), Bengaluru, rapid erosion of forest cover is bound to have a devastating impact on Kerala[1]. This study indicates three important changes:

(i) Between 1973 and 2016, Kerala lost 906,440 hectares (9064.4 sq.km) of forest land, which is 24% less as compared to what existed in 1973. On the other hand, area under plantation gone up steeply from 1850 sq km in 1973 to 9999 sq km in 2016.

(ii) Another perilous change during this period is the area urbanized - which has shot up rapidly from 95 sq km in 1973 to 4136 sq km in 2016; It is important to remember and concede the fact that the steep increase in urbanization at the cost of forest area indicates more built up area (such as buildings, bridges, roads etc) which in turn contributes to high run off and high intensity flash floods during monsoon months.

(iii) Simultaneously, the open area declined substantially from 6143 sq km in 1973 to 2103 sqkm in 2016.

This study clearly demonstrates the extent to which the nature has been encroached and distorted as consequence of which flash floods occur repeatedly in Kerala.[2]

Another interesting study[3] carried out by Mohan Kumar (2005) establishes the fact that there has been a steady change in the land use pattern in favour of agricultural expansion at the cost of forest degradation. In particular the area under plantation crops such as coconut, rubber, arecanut have gone up dramatically during the period 1955-2000. At the same time area under forest has dropped substantially at the rate of 5000 hectares per year during the past several decades. Thus the study concludes, "The consequences of deforestation, which also has been widespread in the state, include frequent flash floods and landslides, soil erosion, and silting of reservoirs, causing serious ecological and environmental problems and complex feedback effects on agricultural production".

Krishnakumar, A, Revathy Das, and Dhanya.T.Dharan (2017), in their study[4] clearly demonstrate that the land use and land cover change in the high lands of Kerala would prove to have dangerous consequences. Land use change with special reference to forests and paddy wetlands in Neyyar and Karamana river basins of Thiruvananthapuram district during the respective periods of 1905-2000 and 1968-2013 were analyzed through Survey of India toposheets and LISS imageries using Geographical

Information System. The detailed analysis has revealed that the area under forests and paddy drastically decreased from 33% to 7% and 10% to 2% respectively. "Forests and Wetlands referred as respective 'lungs' and 'kidneys' of the landscapes, are continuously exploited unscientifically in search of profits and means of subsistence. These ecosystems play a central role in functioning of the biosphere, provide various environmental services by regulating climate, hydrological and biogeochemical cycles and directly and indirectly contribute to socio-economic development" (p.190).

Secondary Reasons

There are also other reasons for the disaster which could be regarded as secondary reasons. They are discussed below:

(1) **Poor early warning system:** There are reports which suggest that the weather and climate forecasts were inadequate to get one prepared to face the threat of heavy downpour and floods. I cannot accept this as a reason in the internet era where you can get instant information by getting in touch with many international meteorological and climate monitoring agencies such as NOAA, INCOIS, WMO, NASA, BBC etc. Let us see what the forecasts were by the official agencies in India: IMD indicated that they had issued a red alert and orange alert warnings for Aug 13th and 14th (heavy to very heavy rainfall in isolated places) in Pathanamthitta, Alappuzha, Kottayam and Ernakulam, Idukki and Wayanad districts). The CM's claim was that while the IMD forecast for 9 to 15 August indicated only 9.85 cm of average rainfall, the actual was much more at 35.22 cm. But the key question is whether the State was prepared for the Red Alert?

(2) **Bad reservoir operations:** Seemingly, this is an immediate reason. All the major dams were opened at a time causing severe flood and inundation in the downstream. What was worse, when the flood gates of the major reservoirs were abruptly opened, the official agencies / reservoir managers had no idea of its impact on the downstream. Had the authorities carried out risks and vulnerability and impact analysis of such abrupt dam opening, the reservoir operations by the PWD would have been more rational. In the end, the worst affected districts were Idukki, Ernakulam, Thrissur, Alappuzha, Kuttanad and Wayanad. Idukki dam created the maximum damage. Apparently, the

operational rules for dam opening during critical periods do not seem to exist.

(3) **CAG Report on Kerala land and reservoir management:** According to the CAG report, the disaster could have been prevented but for the poorly managed reservoirs ((https://cag.gov.in/content/land-management-kerala). The CAG also blames on the ineffective and outdated forecast system. The CAG indicates that such an occurrence took place not only in Kerala but in many other states as well. All dam managers would like to fill their reservoirs during the monsoon months and use the water till the onset of the next monsoon. In August 2014, both the Central Water Commission and the India Meteorological Department had forecast heavy rainfall upstream of the Hirakud dam, on the Mahanadi in Odisha. But this warning was not taken seriously and the dam authorities maintained the water level much above the safe limit with a view not to risk scarcity during the following summer. But the government paid huge price due to abrupt opening of the floodgates of the Hirakud dam. Similar stories could be heard from Gujarat and Rajasthan (2017) and Chennai (2015).

(4) **The role of multi-purpose dam projects:** What is critical to remember and concede is the fact that dams are multi-purpose projects such as storage of water, hydro power generation and flood control (and moderation). But, the PWD tends to forget the last one. In the case of Kerala, as the dams were getting filled the officials were indeed happy that their dams were getting filled. Yes, it would have been happy news during the normal year. But unfortunately, the officials failed to carry out the correct calculations despite orange and red alerts from IMD. In fact, the officials must have thought that after all, the Idukki dam was getting filled up after a span 25 years.

(5) **The dangerous miscalculation:** The cost of paying towards disaster if the reservoirs are abruptly opened is much more than the cost that the state would be paying in case of storage is less than FRL in the reservoirs. This is the lesson which all the states should learn. The question is whether the wisdom would prevail.

(6) **Raising population density:** Population density in Kerala is the third highest in the country (859 per sq km) next only to Bihar (1102) and West Bengal (1029). Even in Tamil Nadu, which is

the most urbanized state in the country, the population density is only 555 per sq km. This clearly shows that the Kerala state is more land scarce: This may be partly because, over 50% of the geographical area in Kerala state is classified as forests of several kinds (including plantations). Still population per kilo meter of forest area in Kerala is far better (1699) compared to many major states barring the north eastern states.

(7) **Declining land space:** Encroachments on the hills and mountains, promotion of tourism in the prohibited area, quarrying and other commercial activities are regarded as the principal reasons for flood damage. Extensive encroachments in the floodplains of rivers have created havoc. In particular, Munnar, Wayanad and Idukki areas and close to water spread areas of water bodies including dams which are regarded as the instant reasons for severe damage; We call it a disaster!

(8) **Official disturbing policy:** The draft policy on scientific mining released by the State Planning Board proposes to encourage geo-mapping of mineral resources in the state and minimize environmental damage. The new Scientific Mining Policy describes itself as a three legged stool. Leg one is uninterrupted supply of extracted natural materials like mineral, rock and value added products - the essential inputs in mineral based industries and in keeping and making modern infra-structure. The second leg is minimizing the environmental damage, while the third one is minimizing the social impact of mining and quarrying operations. The policy also aims at stimulating a 5 percent in Gross State Domestic Product on account of mining activity by augmenting investments in the sector through private participation (Planning Board of Kerala).[5]

National Mineral Policy 2008 allows a greater participation of private sector. The National Mineral Policy, 2008, declared by the Government of India (GoI) replaces the National Mineral Policy, 1993. The new policy allows for large presence and participation by the private sector in the mining sector in order to augment the contribution of mining sector to the National Gross Domestic Product (GDP) and more employment for the skilled and semi-skilled or unskilled people in the employable age group and secure supply of minable minerals and rocks for the mills, smelters and factories.

The Kerala policy which is in tune with the national policy aimed at a growth of 5% in GSDP on account of mining activity in the State (Kerala Scientific Policy, 2015, Draft, Govt of Kerala).

What is flood?

In the context of increasing flood related disasters all across South Asia, understanding flood is important. Flood is a hydrological concept. Stagnant water is not a flood. A pond or a lake when it stores water we call it Pond or lake. But if the same lake or pond receives water after it is encroached for purposes of constructing industries, hotels or houses, we call it flood. Most importantly, flood refers to flowing water. Hydrologically, floods may occur due to over flowing water from water bodies such as rivers, lakes, ocean etc. When water crosses their given boundaries floods occur. Floods can also occur in rivers when the flow rate exceeds the capacity of the river course. Some floods develop slowly, while others such as flash floods, can develop in a quick moment, and even without any visible signs of rain. Therefore, in order not to get caught in the "floods" it is necessary that we respect the water bodies. Encroachments of any kind into the given boundaries of water bodies (including floodplains) should be avoided. In the case of Chennai flood 2015 as well as Kerala flood 2018, it is the clear case of encroachments which caused serious flood devastation.

Impending Loss due to Flood Exposure

The World Resources Institute developed a tool for flood analyzer along with four Dutch research organizations: (https://www.wri.org/blog/2015/03/world%E2%80%99s-15-countries-most-people-exposed-river-floods). The flood analyzer estimates realistic percentage of country's GDP loss due to exposure to flood. Floods in 164 countries have been analyzed and ranked and the top 15 countries accounting for 80% of flood damages have been identified. All the 15 countries are least or less developed countries. The GDP exposed to flood is the highest in India (4.84%), followed by Bangladesh (3.48%) and China (3.28%). Indeed, the extent of GDP exposed to floods in future (in the course of next 50 years) will be enormously high due to utter lack of preparedness and unscientific approach, massive development programmes induced by extensive infrastructural developments, urbanization and industrial programmes. The added threat is the global warming induced climate change which may increase the frequency of occurrence as well as the degree of intensity of floods. In such an event, more the development, the more intensive and extensive would be flood damage. "Using a middle-of-the road scenario, the Analyzer estimates that India's current $14 billion in GDP exposed annually could increase more than 10-fold to $154 billion in 2030. Approximately 60 percent of that increase could be caused by socio-economic development".[6]

Disaster Management - Flaws

Disaster management simply means rescue, relief and rehabilitation. The disaster management policy revolves around these three issues and never goes beyond. Yes, they are important. But lessons learnt from the past extreme events are more important in order to minimize the losses in all future extreme events. Furthermore, it is extremely necessary to concede the fact that the state has encouraged in the past couple of decades mindless expansion of mining and quarrying, tourism, housing development and so on in the areas vulnerable to flood and landslides. The key questions are, is it possible to reverse such expansion and developments? If not, will the state prevent at least such developments and expansions in future?

What is most important is to carry out a risks and vulnerability mapping in case of contingencies such as the one which many states have experienced (the last minute opening of the reservoirs). If such a real time risks and vulnerability mapping is prepared, the PWD will not dare to open the reservoir in the most critical final hours.

Building Resilience

What is the biggest hurdle in building resilient communities in countries such as India where at least one-third of the population live in poverty or below poverty line (that is 430 million people)? This is much higher than the US population in 2018 which is 327 million and more than half of entire Europe's population in 2018 which is 742 million. Most of those who occupy vulnerable areas such as flood plains, river banks, low lying areas such as water sheds etc are the poor and homeless who migrate to cities on the hope for a better survival conditions. In effect, rural poverty is converted into urban poverty and contributes enormously to the pressure on urban infrastructure. In addition, the governments and the "markets" (in mutual collaboration with each other) occupy such public spaces for promotion of the so called development. As a measure of building resilience it is important that this process is stopped. Increasing demographic pressure and population density are biggest concerns. And, it is very critical that we at least comprehend now that higher economic growth and damages to ecology and environment are positively associated and so also directly related to increased exposure to disasters such as floods, landslides and droughts.

Pathways for Way Forward

- First of all, one should recognize and concede that the Welfare State in a democratic governance is the *protector* of ecology and environment (and, all other natural capital) rather than a mere provider under the public doctrine principle.
- It is very important that each State learns from the past events of extreme events and prepare a *realistic* pathway for *maintainable* growth and development.
- Being the monsoon dependent agriculture, Indian farmers are going to get affected very seriously due to temperature rise of 1.5 Degree Celsius (plus). As a consequence, more intense extreme events such as droughts and floods could be expected. Under these conditions, the need of the hour is to frame a more proactive climate policy and so also for the protection of natural capital, in particular high lands and the climate sensitive Western Ghats.
- In addition to promulgation of policies, the State should reduce or *eliminate the gap* between policy and practice.
- Need more effective *law enforcement and monitoring mechanisms* rather than new laws.
- It is extremely vital to build a *scientific and credible data base* on long-term temperature and rain fall variations, on land use and land cover changes, long term changes in soil moisture conditions, long-term changes in bio-mas production, frequency and degree of intensity of occurrence of droughts and floods and long-term changes in agricultural practices and income levels of farmers. The State, research organizations / Universities and NGOs should be more active in creating *Data Bank* on all these important "climate and climate stress indicators" and "natural capital".
- More proactive policies are needed to secure the livelihoods of rural and urban poor.
- For local problems, solutions should emerge from below and cannot be imposed from above and one should go beyond mere expressions of normative concerns.

Concluding Observations

And, do we ever care for our nature? We develop some momentary love and attachment to nature only when we encounter disasters such as high intensity droughts, floods and landslides. Nonetheless, very rapidly we forget and learn no lessons from the past events. It is very critical that we concede that distorting nature would contribute to more disastrous extreme events. And it is also very critical for us to recognize that forests and overall vegetative cover, soil structure and thick soil moisture conditions, watersheds and water bodies are incredible carbon sinks, the loss of which would contribute to increase in carbon foot print and rise in temperature. The worst affected people (as we see today) under these failing conditions are the vast majority of rural and urban poor (the marginalized sections of the society). The most affected rural poor, eventually migrate to urban areas, converting rural poverty into urban poverty. And, unfortunately they are also blamed for expanding slums, encroachments and for adding pressure on the urban infrastructure. Reversing this trend is challenging but necessary.

Notes

1. Four Decades of Forest Loss - Droughts in Kerala, Energy and Wetland Research Group, Centre for Ecological Sciences, Indian Institute of Science, Bangalore; http://wgbis.ces.iisc.ernet.in/energy/.
2. See also, Ramkrishnan, R and T V Ramachandra, *Four Decades of Forest Loss: Droughts in Kerala,* Energy and Wetland Research Group, Centre for Ecological Sciences,Indian Institute of Science, Bangalore; Web: Url: http://wgbis.ces.iisc.ernet.in/energy/(Poster Presentation)
3. Mohan Kumar, B (2005): Land use in Kerala: Changing scenarios and shifting paradigms, in *Journal of Tropical Agriculture* 421(1):1-12. January 2005
4. "Land Cover Change Analysis with Special Reference to Forests and Paddy Wetlands of Neyyar and Karamana River Basins, Kerala, SW India Using GIS and Remote Sensing", in. *International Journal of Scientific and Research Publications*, Volume 7, Issue 11, November 2017
5. http://www.indiaenvironmentportal.org.in/content/413354/kerala-scientific-mining-policy-2015-draft/
6. (https://www.wri.org/blog/2015/03/world%E2%80%99s-15-countries-most-people-exposed-river-floods).

12

Recovery and Reconstruction after the Flood: Challenges from the Health Sector

V. Ramankutty

Introduction

The floods that hit Kerala in August 2018, estimated to be among the most damaging in almost a century, have resulted in major loss of life and economic assets in the state. The government and the people have geared up for a long process of recovery and reconstruction ahead. All sectors have been affected, and health has been no exception.

Consequences in the health sector belong to broadly three categories: immediate loss of life due to the floods, infections and results of displacement in the post-catastrophe period, and long term damage including destruction of health facilities which will affect the health of the population.

Kerala has been known for its impressive achievements in the health sector; therefore any set back in this sector is of special concern to the state. The natural calamity affects health at several levels, and therefore the rebuilding of the sector also has to be planned along those lines.

The Extent and Nature of the Damage

Officially, loss of life due to the floods has been pegged at less than 500, which in itself is not a huge number in a state of around thirty three million people. However, many studies in other parts of the world have pointed out that to understand the true extent of loss of life due to a catastrophe, we need to estimate the excess deaths after the event. Recently the Trump administration in the US claimed that loss of life was minimal after the 2017 weather catastrophe that affected Puerto Rico, which is a

US territory. However, scientists have challenged this claim in various fora, and pointed out that many lives lost due to inadequate health care, loss of access to drugs in time, and other such events would not have happened without the event. This is still being debated. People may die due to causes attributable directly to the floods, and we may not be aware of this. This includes the loss of life due to disruption of care in the case of people on long term support such as dialysis and ventilators, or people immobilised due to stroke. They may not get the continued care after the floods and their lives may be cut short because of this. Many people on long term drugs for ailments such as hypertension and diabetes may have difficulties with continued regular medication during and in the immediate post flood period. The loss of life due to these causes can only be estimated after the event by measuring the immediate post catastrophe mortality and comparing it with the expected rates, an exercise which can be undertaken only after at least one year of the flood. But this is something that must be done.

The actual loss of life attributed directly to the floods has been kept at a low level due to remarkable relief efforts, but many deaths in the immediate post flood periods could be traced to the effect of the flood. There have been reports of suicides and many of the heart attacks and strokes immediately after the floods could have been precipitated by the tragedy afflicted on families. The disruption of water supply and contamination of water sources, coupled with the total destruction of sanitary systems in many parts of the state, will add further to the burden of disease in the immediate post flood period in the form of diarrhoea, dysentery, leptospirosis, and other waterborne diseases. People at the extremes of age like childhood and old age, deprived of their homes and forced to live in overcrowded camps, will be further prone to other infections. But current reports indicate that these have also been contained to a large extent. This is certainly due to the timely and effective action of the health services department, and in spite of adverse propaganda from many quarters against some interventions.

The threat comes not only from infectious diseases, but the disruption of life and the break in regular medication and care that it brings in is sure to affect the course of non-communicable diseases also. It is well established that the major cause of death in Kerala is cardiovascular disease including diabetes and hypertension, accounting for about 40% of deaths. There are reports of disruption of services such as dialysis which cannot but result in adverse consequences. We should undertake an audit of the damage brought on by this disruption.

The floods have also damaged, sometimes beyond repair, many of the health facilities in the state such as sub-centres and primary health centres. While many have been rebuilt, repair and reconstruction drains scarce resources from the state's kitty, resources which could have been utilised for more constructive purposes. The state is in the process of launching a major initiative in primary health care, the 'Aardram' programme, which will strengthen many sub-centres and PHCs in the state. The extent of setback which the floods have brought in is not yet assessed.

How to Address this Challenge?

Nevertheless, it has also been claimed that compared to such departments like revenue and transport, perhaps the health services department has responded in a very timely fashion to the sudden and unexpected catastrophe. This could be because they have a network of information systems, used to conveying information on outbreaks of epidemics immediately to the decision makers in the capital. It is true that this machinery swung into action and was in full flow within hours of the disaster setting in. It also incorporated many NGOs and voluntary groups focussing on various aspects, and even institutions like the National Institute of Mental Health and Neurosciences Bengaluru, who had a field team ready to take care of the distressed people affected by the floods. Paradoxically, it has also been claimed that since the health department acted promptly and prevented early deaths, there is a false sense among the authorities especially at the Central level that the damage due to the floods has not been too heavy.

In addressing this multiple challenge, the state has some strengths which it should draw on. First is the strong public health system that the state has built up over the years. The information network that the health services have built, and its recent modernisation, is a very strong support tool for the recovery and rehabilitation efforts. The decentralisation of decision making in health which can be traced back to the initiatives at Panchayat level planning twenty years ago has given the local governments sufficient experience in initiating new programmes at the local level. This was evident in the way the local representatives took charge and worked day and night in the flood recovery efforts, along with voluntary people from the groups. The highly educated population which is sensitive to the media and willing to cooperate with civil society groups is another advantage that the state has. What the disaster brought out was that contrary to what many people believe, young people in the state responded to these distress calls by volunteering their labour and time, and their expertise in using the new media tools is very handy in such situations. In many situations, more than the the official machinery, it is such groups that acted first. Moreover, it was evident that they were very much willing to work with the official teams. However, since the floods have receded, there is no longer a glue which keeps them together, and the team spirit seems to have been dissipated.

Perhaps the greatest challenge to the state is mobilising the resources needed; not much support is expected from the Central government in this regard. However the state is very much capable of generating the resources. The policy makers in charge should think of imaginative plans to raise resources and not the kind of effort that has the potential of putting off sections of the people. The resource generation should also be a transparent and accountable process, and not one where unscrupulous elements have the opportunity to amass wealth, a phenomenon that is all too familiar to the people in the state.

Local problems are best assessed by LSGs and they can formulate plans to tide over the crisis. The state government needs to divert funds for this reconstruction in the health sector. Local NGOs need to work with the LSGs in mobilising support and extending their reach to marginalised sectors such as tribals and coastal communities. Rebuilding health facilities which once offered services to the people is a huge financial challenge; but it is also an opportunity to get rid of unusable buildings and equipment which is not working. This has been demonstrated by many health facilities which have been rebuilt in the short time since the floods.

Conclusion

All these advantages should be harnessed in addressing the challenges in the health sector. This is where the leadership plays a key role. They should be ready to engage with people's initiatives and not be confined to technical approaches. This setback has come at a time when the health services department was poised to make a major break through with its initiative to innovate the health services in the state. This should be taken as an opportunity to completely reorient the health system to one that is more sensitive to the people's needs and based on the principles of primary health care.

13

Rebuilding of Kerala through Ecological Civilization

V.S. Vijayan

Background

One of the most irreversible damage caused to man and nature is the Industrial Revolution; a revolution based on fossil fuel energy (such as coal, natural gas and petroleum) which ultimately created a centralized platform for development giving way for accumulation of wealth at the expense of nature, consequently severing the very umbilical cord between man and nature.

In 1858, Karl Marx, following Fraas, wrote: "Cultivation when it proceeds in natural growth and is not consciously controlled leaves deserts behind it". Subsequently Marx and Frederick Engels wrote on the environmental issues like climate, soil degradation, air and water pollution, over exploitation of natural resources, over population, deforestation, desertification, industrial toxins, and the destruction of species and ecosystems. Significance of this statement is that all these problems that the world faces today at an alarming rate came almost a century after the Industrial Revolution of 1750s.

Al Gorey in 2006 in his Oscar winning film the "Inconvenient Truth" reveals *inter-alia* how higher is the carbon dioxide presently in the atmosphere than in the past 6,50,000 years, using *Antarctic ice coring* data (*"Annual Mean Carbon Dioxide Data", NOAA/ESRL; Amos, Jonathan (2006-09-04); "Deep ice tells long climate story", BBC News)*. Carbon dioxide level increased from pre-industrial levels of 280 parts per million by volume (ppmv) to about 396 ppmv in 2013 (Environmental Protection Agency, USA, 2015).

The inevitable need for a brake in the onward march of industrialization was, probably, brought in by Rachel Carson in 1962 through her much acclaimed sordid narration of the use of chemical pesticides and how they kill the man and nature.

Of course development civilization at the expense of nature has certainly created wealth (in a few hands) and gave 'comfort' for a few. In India in 2012, we had 1.53 lakh millionaires and 59 billionaires, whereas in 2017 it has gone up to 2.45 lakh and 136 respectively. Automobile industry has certainly registered an impressive growth; the number of cars shooting from 3 lakh to 400 lakh between 1950 and 2010.

Sky scrapers, multi-storeyed malls challenging the West, huge mansions, flyovers, metros, highways, hospitals, industries of various hues and, all such infrastructures have increased incredibly, mostly at the cost of non-renewable resources polluting our environment and food. GDP grew up, from 3% of the 1950 to 9.59% in 2009. Although it crashed to 4.74% in 2014, all out attempts are being made to bring it up to 8%, even higher, in 2020.

While these are the shining part of the development civilization in India, the other side of India is mired in poverty, slums, and filth from waste mountains, pollution, suicide, rape and a seemingly incorrigible corrupt system.

According to 2011 census data, child death before the age of five every year was 21 lakh due to diarrhoea, typhoid, malaria, measles, pneumonia; slum dwellers about 4.26 crore; homeless 2.76 crore; people without a toilet is 70% (PRI's the World, 2016);40% children under the age of 5 are stunted (Water Aid Report 2015),17 million with moderate malnutrition (India Health (Report 2016); undernourished population in India was 250 million (FAO, 2012) and, suicide during 2005 - 2012 was 2, 84,694 (National Crime Records Bureau).

These apart, was the heavy toll on the agro-biodiversity; India had around 1,10,000 varieties of rice. Only a few (around 1500?) are left currently in pockets in North-East and in tribal colonies where green revolution has not reached. Fortunately large scale cultivation is confined hardly to 30 hybrid varieties. Forests have dwindled. In Kerala, Tamil Nadu and Karnataka alone, 40% of the forest was lost during the period from 1920 to 1990, due to one or more of the factors such as mines, quarries, illegal constructions, encroachments and tree felling. All these lead to climate change, water shortage, soil erosion, dying of rivers etc.

Wetland scenario is still worse. In India, 38% of wetlands were lost within 10 years from 1991 to 2001 (Vijayan *et al.*, 2004). In Kerala alone, the State had around 7,53,386 ha of wetlands in 2004 (Mohanan, 2004, Govt. of Kerala, 2013), whereas it has gone down to 3,71,185 ha in 2011 (MoEF, 2011; Govt. of Kerala, 2013). The loss is 3,82,201 ha (51%); that too within 7 years. The loss of Paddylands between 1975 and 2017 is 7.00 lakh ha, totalling about 10, 82,201 ha.

Pollution level is appallingly high. A study of 1700 fishes from 140 wetlands from 14 States showed none was free from either heavy metals or pesticides (Vijayan *et al.*, 2004). 60 percent of India's water-sources (which are routinely monitored) have poor "bio-chemical oxygen demand", an indicator of organic pollution; about 68 percent have faecal coliform; 79 percent metropolitan cities had very high levels of particulate matter and nitrogen dioxide, causing air pollution (Annual Report 2011-12, Central Pollution Control Board).

The most disastrous consequence of development paradigm that we have been following is that, today mother's milk is polluted, embryo in the mother's womb has toxins and, the umbilical cord of the new born baby has about 287 chemicals of which 180 are carcinogenic (9 December 2009; Environmental Working Group USA). Altogether, 493 different kinds of chemicals were located from baby to grandfather.

Therefore, it is in this background one should think of rebuilding Kerala which was devastated by the floods and landslides. The question is; should we continue to have this kind of fossil fuel based GDP oriented development paradigm ravaging the natural resources, polluting the air, water, soil, and food resulting mainly in the creation of a few rich while pushing the life of majority in to appalling misery.

Basic Ecology of Kerala

Kerala's ecology is dependent mainly on two physical features, namely the ecology of the Western Ghats which run across the length of Kerala except a gap in Palakkad (remember that is why Palakkad is facing higher temperature and less rainfall compared to other parts of Kerala), and the 41 rivers that originate from the Ghats and emptying in the Arabian Sea, although with enormous man-made blocks and hindrances on the way.

Therefore, the Western Ghats (WG) and these rivers and their flood plains, wetlands (ponds, tanks, canals, streams and rivulets) have to be given special attention for building a resilient Kerala.

A few most essential vital points for the sustainable use of the resources of the WG and the wetlands in the State - the life-line of Kerala - and the related areas are listed below. Please note that the Sálim Ali Foundation does not stand against development, we insist that development should be sustainable and should not be at the expense of the total destruction of our forests, wetlands and environment.

Major Policies to be Adopted for Rebuilding

At the outset six major policies are required for the protection of environment and ecological security of the State.

1. Hill Conservation Policy, Strategy, Action Plan and an Act to implement the same. This could be worked out with the participation of local communities, apart from experts. Time: 6 months
2. Land Use Policy: A draft of the Land use Policy was submitted to the Government in 2009 by the Kerala State Biodiversity Board. A Committee Chaired by the then Chairman, Kerala State Biodiversity Board (Dr. V.S.Vijayan) was appointed by the Revenue Department for formulating the policy. This may be reviewed and implemented within 6 months. Once this is in place, most environmental problems will have solutions.
3. Kerala State Wetlands Wise-use Policy, Strategy and Action Plan to be formulated within six months with public participation. Necessary acts and rules to implement the same may also have to be formulated.
4. Kerala State Sustainable Development Policy, Strategy and Action Plan to be formulated and implemented.
5. Apart from these, the existing Kerala State Organic Farming policy, Strategy and Action Plan adopted in 2010 has to be taken up seriously and implemented with will and determination.

The most vital point is a determined shift from the present development paradigm which destroys all our natural resources to a sustainable development paradigm conserving our natural resources to the maximum extent possible.

Action required in each area is given in the following sections

Western Ghats

I. Mining & Quarrying

1. In the Western Ghats (WG), in Zone 3 new mining can be permitted but only for scarce minerals not available in the plains with strict social audit and EIA. (ESZ, Ecologically Sensitive Zone 1,2,3 as described by the Western Ghats Ecology Expert Panel).
2. The existing quarries in Zone 1 & 2 shall be banned immediately. And, no new licences shall be issued.
3. Quarrying licence in Zone 3 can be given but after a review, case by case and Cumulative Impact Assessment Studies.
4. Ban all illegal stone quarries immediately, which appear to be more than half of the existing quarries.

II. Land Use in the WG

1. Change of land use from forest to non-forest use or agriculture to non-agriculture shall not be permitted, except for agriculture to forest, and, except when extension of existing village settlement areas to accommodate increase in population of local residents.
2. Respecting the commitment made at the all party meeting and the Government's decision in 1992 for regularising the land encroached up to 1.1.1977 shall be implemented after careful examination of the revenue records. This should be done very carefully as there are already umpteen allegations of irregularities.
3. In any case, encroachments on the slopes having 30° or more and those blocking the water course must be evicted.
4. In the case of Sabarimala, there is a dire need for regulating the annual number of pilgrims in a year as the area cannot probably support more than a lakh in a year.

New Buildings and Roads in the WG

1. New railway lines and major roads shall not be allowed in Zone 1 & 2 of the WG, especially at the expense of forests, grass lands and wetlands. However, in Zone 3 such constructions can be allowed subject to the regulations of the government and social audit.
2. New hotels and resorts shall not be constructed on the hill sides or water courses blocking water flow.

3. They shall not be at the cost of forests and wetlands.
4. In any case, high rise buildings shall not be allowed anywhere in the WG.
5. Roads and other infrastructural expansion plans must have EIA scrutiny, assessing especially ecological costs and public benefits.
6. Develop regional systems for handling hazardous, toxic, biomedical wastes as well as recyclable wastes. Local authorities should be made responsible and accountable for these.

Industries in the WG

1. New hazardous or toxic waste processing units shall not be allowed in Zone 1 & 2; existing ones should be phased out within six months.
2. New polluting (red and orange category) industries shall not be allowed in Zone 1 & 2; existing industries should switch over to zero pollution by 2020 and be subject to strict regulation and social audit.
3. All the industries situated on the bank of rivers should be made pollution free within a year, otherwise they should be closed down. Licence for new industries can be given only after confirming that they are pollution-free.

III. Power/Energy

1. Run of the river schemes can be taken up, after Cumulative Environmental Impact Assessment (CEIA) with maximum 3 m height to serve local energy.
2. No new hydropower project can be permitted anywhere in the WG.
3. Promote solar / biomass sources for decentralized energy needs.
4. Dams and thermal projects that have crossed their viable life span (for dams the threshold is 30-50 years), outlived their utility, underperforming, silted up beyond acceptable standards, etc. to be decommissioned in phased manner.
5. Please note that decommission is not demolition. Both are quite different. For the former, we have to follow the guidelines given by the International Commission for Dam Decommissioning.

Process involved in decommission of a dam are assessment of: *(a) Structural safety* due to age, weakness, structural problems, construction defects, *(b) Reservoir siltation:* inability to store water as envisaged, *(c) Reduction in benefits* due to poor design, inefficient turbines, *(d) Economic costs:* a balancing of the maintenance cost and the benefits that the society currently get, *(e) Ecological damage* caused to biodiversity and other economic loss. All these require in-depth studies by expert committees in the related areas, and finally the government has to take a call on this by assessing all the reports. It is not just demolition of the dam as many people are worried about.

1. Special Economic Zones and new hill stations shall not be allowed in the WG considering that the entire area is already saturated and it cannot take any further burden as it has reached its carrying capacity.
2. Alternatives for plastic bags should be made cheaper and popular and, production of plastic bags stopped and use of plastic bags banned within a year.
3. Conversion of public lands to private lands shall not be allowed.

IV. Forests

1. Water conservation should be the most important objective of the forest management. Accordingly evolve and implement an ecosystem and landscape based approach for forest management.
2. Ensure watershed and biodiversity conservation in the catchments. Afforestation of catchments of dams and rivers should get priority.
3. The planters holding extensive lands under coffee, tea, rubber, cardamom in the high ranges and high altitude areas shall carry out assisted eco-restoration of catchments so as to improve flows and heal land degradation in a phased manner.
4. Ensure one-third of the land area of the State is kept under forest (excluding plantations of all types) as envisioned in the National Forest policy, 1988.
5. The existing forests shall be conserved in totality and no forest shall be de-notified for any schemes including laying roads, irrespective of their public appeal.
6. However, in the least important ecologically sensitive areas, as suggested by the WGEEP (Western Ghats Ecology Expert Panel), laying of roads can be considered for unavoidable public interests, but only after Environmental Impact Assessment.

V. Wetlands and Paddy lands

Post flood disaster rebuilding of Kerala should treat conservation and sustainable use of wetlands more seriously than ever, because of the bitter experience and lessons from the disaster. Had we protected the wetlands and paddy lands, the magnitude of the flood disaster would have been reduced to at least 50%. By reducing the wetlands and paddy lands (*Wetland loss in Kerala in 7 years between 2004 and 2011 is* 3,82,201 *ha and, paddy lands between 1975 and 2017 is 7.00 lakh ha)* totalling about 10, 82,201 ha, the State has lost that much area for holding water during rains and floods. It may be remembered that the only natural features that can hold water during flood and rain are the wetlands and forests. Therefore, it is imperative to make sure that these vital features be considered seriously while planning rebuilding of Kerala. A few most indispensable points are given below:

1. Formulate a Kerala State Wetland Conservation and Wise-Use Policy, Strategy and Action Plan and make necessary changes in the existing rules to implement it.
2. Ban reclaiming wetlands any further except purely for public purpose fully supported by the Government or public institutions of the Government with the full approval of the LSG.
3. In the Amended Act, 2018 of the Kerala State Conservation of Paddy land and Wetland Act 2008, those provisions which lead to further loss of wetlands and paddy lands should be amended further to make sure that wetlands and paddy lands will be completely protected.
4. In case of absolutely essential roads across the wetlands, consider alternatives such as fly overs without constructing bunds and sluice gates.
5. Any project executed before 2008 but had to be stopped since the Act came into existence, allow completion of those without expanding it.
6. Restore all wetlands reclaimed since 2008 after the State paddy land and wetland conservation Act came into existence.
7. Encourage sustainable utilization of wetlands by participation of local communities ensuring economic benefits to the

people who live in the surroundings as demonstrated by the Biodiversity Board in 2009-2010 in Veliyankode Panchayat in Malappuram.

8. Create a Wetland Register for each Panchayat. Such a register should be with the Panchayat which should be made responsible and accountable for the conservation of wetlands in the respective Panchayats.
9. Complete the work of data bank on a war-footing within another six months.
10. At no cost shall the paddy lands be reclaimed for any other purpose, including expansion of highways. In most essential areas, resort to fly-overs without filling paddy lands or wetlands can be considered.
11. It must be born in mind that the paddy lands, apart from using for cultivation of paddy, can be used for fisheries and poultry (duck) as already demonstrated elsewhere in the country.
12. Apart from these benefits, wetlands and paddy lands serve as a source for recharging wells and ponds which no other natural feature can do, except of course rain which is unpredictable quite often.
13. It may also be noted that the ecosystem service values of wetlands and paddy lands is Rs.90.1 lakh per ha, that too every year. The values of all types of wetlands and paddy lands in the state put together is Rs. 2.75 lakh cr. every year, more than double this year's budget!
14. Therefore, it is more than justifiable that the State give a portion of these values to the owners of wetlands and paddy lands as a reward for keeping them. This will encourage the owners to preserve them rather than selling or reclaiming.

VI. Agriculture

1. Every inch of cultivable land in the State should be cultivated.
2. All fallow land should be converted for cultivation, either for paddy or vegetables depending on the local conditions and water availability.
3. Switch over to organic farming and make sure that food crops are converted into organic within 3 years and the cash crops within another 3 years.

4. This would be possible, only if a Mission for Organic Farming as envisaged in the Organic farming policy of the State is set up with a time targeted mission to work independently but in tandem with the Agriculture Department.
5. Tapioca and other tubers shall not be encouraged on steep slopes (>30°), as when they are harvested, the soil get loose and washed off during rain. Farmers who shift such cultivation to plain land should be given financial compensation, especially for small landholders.
6. Must evolve a time bound programme to make the State self-reliant in vegetable production.
7. Set up a *Department for Urban Agriculture* focusing on agriculture development in the urban areas.
8. All the houses must have some cultivation of vegetables according to the availability of land including space on the terrace.
9. Houses with large lawns and flower gardens must have vegetable cultivation at least in 50 % of the area.
10. Provide compensation to the farmers who switch over to organic cultivation during the first one or two years when the production is likely to be reduced.
11. Subsidies intended for the fertilizers be given to the organic farmers to help in the organic farming.
12. Take a strong decision that production within and import of chemical fertilizers and pesticides from outside the State should be reduced in line with the expansion of organic farming area in the State.
13. Municipal/corporation waste (to be called resources for organic manure) be separated at source and distributed to villages free of cost for production of organic manure.
14. Educational centres, starting from schools must cultivate vegetables/paddy according to the space availability.
15. Make sure that the water channels be cleaned, blockades cleared for free flow of water. This should be the prime responsibility of the LSGs. This would do enormous help to the farmers.
16. Insist on using traditional varieties of paddy, vegetables and fruits and give financial support in the form of rewards for such farmers. It is to be noted that in the face of climate change, only traditional variety will have the resistance power to cope up with which they gained over decades.

17. Protect all the remaining pokkali fields for sustainable integrated farming, as these are the areas used for eco-friendly rice - fish culture.
18. Provide a small amount as a monthly support to the farmers who practically live on farming which depend heavily on weather which is quite unpredictable, of late.
19. Provide a certain percent of the annual ecosystem services of paddy land to the owners as a reward for conserving the wetlands, as the benefits are drawn by the society.
20. Declare paddy lands as "Paddy Reserves" without affecting the ownership of the land.
21. Write off the loans taken by the farmers affected by the deluge and landslides
22. Genetically Modified Crops shall not be allowed anywhere in the Western Ghats and other parts of Kerala.

VII. Coastal Ecosystem Management

1. The State Coastal Zone Management Authority should take stringent action against the violators and has to take effective measures to monitor and check violations.
2. The functioning of the State Coastal Zone Management Authority has to be decentralized by establishing District level authorities with adequate legislative and financial powers to monitor and to implement the Act.
3. The reclamation of lakes and coastal wetlands should be banned totally and the existing boundaries are to be identified and fixed using advanced technologies including satellite mapping to prevent further encroachment.
4. Strict guidelines shall be formulated to control mining of strategically important heavy minerals, namely black sand (such as limonite, monazite and rutile) and industrially important minerals, namely glass sand (white sand - silicates) and, the same may be enforced strictly.
5. To prevent human settlement in CRZ, separate fishing townships with all facilities required for the fishing communities should be developed outside the No Development Zone of the Coastal Regulation Zone (CRZ).
6. Notwithstanding the new coastal zone notification reducing the no-development zone to 50 meter, if the Coastal Zone have to be

protected, a minimum of 200 meter from the HTL is to be earmarked as no-development zones. This area should not be utilized for any other purpose, except for fishery related activities of the coastal fisher folk.

7. No reclamation or constructions after 1991 should be regularized. This has particular relevance where industrial clearances are sought from the MoEF.
8. Restoration of beaches and protection of sea shore through natural shields such as mangroves and typical coastal vegetation have to be promoted with the participation of coastal communities.
9. In no case shall granite walls be constructed to protect the shoreline, which is ineffective and cause immense damage to the already fragile WG ecosystem.

VIII. Pollution Control

1. Industrial effluent disposal in the coastal water bodies has to be curtailed. Existing industries should be made zero pollutant within a year and the new ones must be zero pollutant.
2. Sewage collection and treatment mechanisms have to be made mandatory for house boats operating in the backwaters.
3. Local bodies should take urgent measures to control the solid waste generation and take eco-friendly measures for proper recycling or disposal of the wastes.
4. Beach Clean-up programmes should be initiated with the help of civil society organizations and coastal communities.
5. Coastal tree shield with indigenous coastal vegetation of a minimum of 10 m width should be promoted in areas of human settlements, with the participation of local communities.

IX. Biodiversity Conservation and Sustainable Use

1. By-catch reduction methods should be made mandatory in trawl nets to reduce the loss of biodiversity, especially the destruction of RET (Rare, Endangered, Threatened) species.
2. Principles of Ecosystem Approach to Fisheries (EAF) and Code of Conduct for Responsible Fisheries (CCRF) should be adopted to manage marine fisheries of Kerala coast to sustain the productivity.
3. Special Economic Zones in certain potential fishing areas and fishermen hamlets for complementing high-tech projects should not be encouraged.

X. Combating Climate Change

1. Integrated coastal area development programme covering socio-economic conditions of fisher-folks in coastal areas with participation of the local communities should be launched immediately.
2. Considering the impending sea level rise, adaptation strategies including a proper rehabilitation programme should be in place soon.

XI. For Responsible Tourism

1. The responsible tourism guidelines should be revised, if necessary in the light of the recent flood and landslide, and followed for all the tourism initiatives in the coastal and marine areas.
2. The number of houseboats in backwaters should be strictly limited, based on carrying capacity studies and strict rules and regulations for waste management, especially in backwater tourism using house boats, shall be formulated and enforced.
3. All house boats shall run on non-conventional energy sources such as solar and be fitted with green toilets/safe disposal of the wastes.

XII. Housing

1. Houses should not be allowed on terrains with more than 25° slope on the WG.
2. It shall not be on the water course, blocking the water flow.
3. Buildings shall be sited at a minimum of 100 meter from the rivers.
4. In the case of rivers with dams, the maximum point where the water spread reaches at the time of opening the dams fully, may be marked and no buildings shall be permitted inside the water spread area.
5. The building code of 2005 may be revised with provision to:
   (a) Make sure that all buildings are environment-friendly, using minimum non-renewable material such as granite stones, steel, and sand. It is not to ban constructions of buildings, but to reduce the use of such material.

   (b) All the new houses and the existing ones should have solar panels to meet at least 75% of the power requirement. Government must come forward with attractive schemes to support those belonging to BPL.

(c) Every house should have a roof water harvesting scheme.

(d) Waste management schemes be made mandatory for the flats with segregation at source as organic and inorganic. The former be given to the farmers for production of organic manure in the neighbouring villages.

(e) Collection and dispersal of the waste from the houses and flats should be the responsibility of the ward members and the Panchayats/ Municipalities/ Corporations.

(f) Maximum size of a house should be fixed depending on the number of members in the family. And, the license should be given only if it conforms to the rules.

XIII. Industries

1. Industries of any type shall not be located on the river banks obstructing the water flow.
2. Each district must have a site for industrial development, considering the water availability.
3. Promote industries that do not depend much on ground water.
4. Encourage industries using local resources.
5. All industries must be zero polluting and shall not pollute air, water or food.

XIV. IT hubs

Instead of setting up huge IT centres in one or two cities, causing pressure on housing, water supply and transportation, decentralise them across the State even at block level. This has immense social and environmental advantages.

XV. Transport

1. In the case of roads and railway lines in the mid lands and plains, they should not be at the expense of paddy lands and wetlands.
2. Where the need occurs for expansion of roads at the expense of wetlands or paddy lands, alternatives should be sought of including fly overs.
3. When the need for connecting two villages on either side of a large wetland or paddy lands, a bridge may be constructed on pillars but no bunds with sluice gates should be thought of. Water flow shall not be prevented. We have already paid dearly for such blunders during the recent floods.

4. Roads already in use shall be converted to climate and eco-friendly in a phased manner to provide efficient and safe transportation reducing consumption of fuel, traffic congestion and environmental pollution.
5. All roads under the various panchayats, PWD and NH which are in poor condition should be repaired on a war-footing and made weather proof, and climate resilient to the extent possible, with a guarantee for at least 15 years.
6. There should be safe pedestrian and cycle paths all over the State.
7. Bridges and Culverts in dangerous condition should also be repaired / replaced instead of going in for high investment new corridors and roads.
8. Public transport system should continue to be run by the Public Sector, but should be made more efficient, comfortable, keeping strict time schedule and, should be subsidised, considering the burgeoning fuel cost.
9. The public transport system should be made comfortable and cheaper and the roads laid and maintained.
10. Renewable energy and non-polluting transport systems should be introduced, especially CNG (Compressed Natural Gas) for vehicles in a phased manner. Strict enforcement of pollution control norms is needed.
11. While the existing roads through the forests, wetlands and mangroves may be maintained well, no new roads or rail lines shall be permitted at the cost of these valuable ecosystems.

These points if implemented, Kerala will become the first in the country adopting an ecological civilization and naturally on its way to achieve sustainable development. In such an event, the impact of calamities would be the minimum.

14

Enabling Environment for Climate Smart Agriculture

Saji John

1. Introduction

The agricultural sector has witnessed its most challenging and taxing times, after the devastating floods that wreaked havoc on the State of Kerala. Among many affected, the agrarian population is facing the worst adverse effects of flooding and soil erosion. The scientific community has emphasized on the urgent need for reclaiming the eroded soil to restore its fertility, glorious biodiversity and capacity to sustain all life forms. While considering the prospects of rehabilitation, it is equally important to realize that the prime cause of such calamities are recurring aberrations in the normal weather and climatic conditions of the State. Hence, it is worth discussing the Climate Change that has profound influence in the agrarian sector.

There has been projection in a recent report by the Asian Development Bank (*ADB, 2017. A Region at Risk: The Human Dimensions of Climate Change in Asia and The Pacific)* that East, South, and Southeast Asian regions are extremely vulnerable to the effects of environmental change. State of the art models have projected pronounced increase in frequency and intensity of heavy rainfall events in the region which may cause more severe flooding if the global temperature continues to rise.

2. Climate Change

Change is inevitable in everything. Climate Change is a reality of our world, and it has large implications for society in general, with a particular impact on agriculture. Models project that mean winter temperatures in India will increase by as much as 3.2°C in the 2050s and 4.5° C by 2080s. Summer temperatures will increase by 2.2° C in the 2050s and 3.2° C in the 2080s. Further, for every degree of global warming, the world is committed to an eventual sea-level rise of over 2.3 meters (*ADB, 2017*).

In the first report on "Impact of climate change in four regions of the country" by the Indian Network for Climate Change Assessment, it has been pointed out that reduced rainfall, increased atmospheric temperature and flooding due to rising sea levels are the climate change scenarios for the Western Ghats and Kerala over the next 20 years.

2.1 Kerala and climate change

In keeping with the clear upward trend in surface air temperature, the mean maximum temperature over Kerala during the last 50 years has risen by about 0.8 °C, the minimum by 0.2 °C and the average by 0.60 °C over (27.3 - 27.9 °Celsius), as per the Indian Meteorological Department. Such warming not only affected the entire Kerala coast, but also led to widening of differences in the maximum and minimum temperatures along the high ranges of Kerala. The changes in thermal and moisture regime over the year have resulted in the shifting of climate in Kerala from B4 to B2 class, moving from wetness to dryness within the humid type of climate.

Temperature data recorded at the Regional Agricultural Research Station, Ambalavayal (Wayanad district) and Cardamom Research Station at Pampadumpara (Idukki district) under the Kerala Agricultural University revealed that the maximum temperature over the high ranges of Kerala had increased by 1.46° C from 1984 to 2009. It also reveals that the effects of global warming and deforestation are felt more across the high ranges of Kerala situated in the Western Ghats, one of the hot spots of bio-diversity.

Under the projected climate change scenario, it is certain that the temperature is likely to increase by 2°C by 2050. The minimum surface air temperature in the Western Ghats region may rise by 2°C to 4.5°C. The average temperature in the region bordering Kerala is likely to rise by 1°C to 3°C. The number of rainy days is likely to decrease along the entire Western Coast, including the Western Ghats.

It is now established that the sea level is on the rise due to global warming and the projected Sea Level Rise (SLR) along Kerala coast on a conservative estimation is about 100 to 200 mm over the next 100 years.

Vulnerability to Sea Level Rise would be alarming to the majority coastal communities which live on sandy coasts, most of which are barrier beaches or spits. Backwater banks, islands and filtration ponds/paddy fields are other sections of the coastal zone which are highly susceptible to Sea Level Rise. If the sea level rises by one metre, 169 sq. km of the coastal region surrounding Kochi will be inundated.

2.2 Kerala, A climate change prone state

There are many reasons for Kerala becoming a *climate change prone* State. High dependency on climate sensitive sectors like agriculture, fisheries and forests, very long coast line prone to severe sea erosion, unprecedented decline in forest resources and biodiversity, continued soil erosion leading to reduced soil quality and low productivity, urban & industrial effluent discharge to rivers, sand mining from the river bed and clay mining from river banks/floodplains, alarming rates of reclamation of wetlands, high density of dams restricting natural river flow, extreme hunger for land for housing and livelihood, unregulated backwater tourism (house boats) and 'eco' tourism in the ecologically fragile zones etc are the major factors which make our state vulnerable to climate change.

High population density and high state of urbanization result in high per capita energy needs and this enhances carbon intensity. Rapidly expanding infrastructure development as well as large proportion of goods and traffic movements increase Green House Gas emissions. The high energy needs consequent to high urbanization in the wake of high population density persuade even our administrators to compromise on environment susceptibility without integrating green technologies in the development process. Financial constraints of the state also restrict the adoption of cleaner technologies instead of polluting ones.

Rubber plantations cover about 18% of the total agricultural land and 11% of the total geographical area of the state and this has posed threat to our food security, biodiversity and balanced ecosystem. Acute food insecurity and import of food grains and vegetables from other states in many ways aggravates the vulnerabilities. The Centre for Water Resources Development and Management (CWRDM) has identified the probable impacts of Climate Change in Kerala as follows:

- Higher temperature, and higher rates of evaporation and transpiration
- Location- specific higher/ lower precipitation and shifting pattern of monsoon

- Increased surface runoff during monsoon causing floods and decreased summer flow
- Soil erosion leading to degradation of soils, siltation of water bodies and canals, and reduction in capacity of reservoirs
- Higher rates of soil moisture depletion and faster dry up of soil
- Higher rates of groundwater depletion and declining water table
- Drying up / shrinking of ponds, tanks , lakes, wells etc
- Increased incidence of droughts, floods and landslides
- Sea level rise, and Coastal erosion
- Salinity intrusion/ ingress into surface and groundwater in coastal areas

The major climate change hotspot districts in Kerala are Alappuzha, Palakkad and the hilly districts of Wayanad and Idukki. Water scarcity, spread of vector and water borne diseases, degradation of forest and biodiversity, decrease in the agricultural production etc. are some common issues which may become severe in the future.

3. Impacts of Climate Change on Agriculture

FAO estimates that agricultural production will have to increase by 60 percent by 2050 to satisfy the expected demands for food and feed. But climate change is emerging as a major threat to agriculture, food security and livelihood of millions of people across the world. It is estimated that in 2030, India will surpass China to become the world's largest populated country. The burgeoning population urges the best out of the crop production. To feed an expanding population, the annual world food production will need to increase by 60 percent over the next three decades.

Climate change is expected to increase temperature and precipitation variability, reduce the predictability of seasonal weather patterns and increase the frequency and intensity of severe weather events, such as floods, cyclones and hurricanes. Some regions are expected to face prolonged drought and water shortages. All these will adversely affect crop production. Changes in crop cultivation suitability and associated agriculture biodiversity, decrease in input use efficiency and prevalence of pests and diseases are some of the major causes of climate change impacts on agriculture.

It also adversely affects the soil processes resulting in soil erosion, soil leaching, soil organic carbon loss, salinization and nutrient loss. Decrease in soil moisture and lowering of water table hasten organic matter decomposition in soil, thereby enhancing emission of green house gases. Changing weather parameters have a detrimental effect on cropping seasons and cropping patterns leading to many crops currently being raised unsuitable for cultivation in the future. Climate Change also causes immense deterioration of forests and biodiversity, affecting the ecosystems supporting sustainable development. Rise in temperature and precipitation increases the relative humidity, paving the way for new pests and diseases, thereby enhancing the risks of farming. Minor pests may turn into major pest's epidemics in various agricultural regions of the state, thereby adversely affecting crop production and productivity. Further, arable land along the coast lines are bound to be reduced with an intrusion of saline water. Coastal erosion and submergence of shorelines could mainly affect agriculture through inundation of low lying lands.

3.1 Climate change and Kerala agriculture

Climate change has its impact on Kerala agriculture in many ways. Cultivated land is declining year after year across the state and production is almost stagnant. With changes in rainfall pattern, drought, increased rate of runoff causing floods and the erratic behaviour of rivers, the conventional agro-climatic zones of the state need to be re-defined.

Increased temperature enhances the metabolic activity of the plant which reduces crop duration and reduces crop yield. With each degree rise in temperature, rice yield would be reduced by 6 %. Thus an increase of temperature by 2°C by 2025 would affect paddy production in Kerala. Salt water intrusion and salination of fertile rice land are becoming a problem in Kuttanad (Alappuzha and Kottayam districts), Pokkali land (Ernakulam and Alappuzha), Kaipad lands (Kannur District) and Kole land (Thrissur and Malappuram) adversely affecting rice production in the State.

Climate projections across the high ranges indicate that the Southwest Monsoon rainfall is likely to decline and surface air temperature and its range are likely to increase. Under such circumstances, there is a threat to thermo sensitive crops like black pepper, cardamom, tea and coffee. Major perceived effects of climate change on mango include early or delayed flowering, multiple reproductive flushes, variations in fruit maturity, abnormal fruit set and transformation of reproductive buds into vegetative ones. Cashew is also highly weather sensitive. Despite advanced technologies in crop production and crop improvement, cashew productivity is declining over Kerala.

3.2 Impact of agriculture on climate change

Apart from analysing the impact of Climate Change on Agriculture, it is also important to learn how agriculture impacts climate change or global warming. The agricultural sectors, which include crop and livestock production, forestry, fisheries and aquaculture, are major contributors to global greenhouse gas emissions. According to FAO estimates, in 2010, emissions from the agriculture, forestry and other land use (AFOLU) sector directly accounted for 22 percent of total global emission.

In Kerala, the emission of green-house gases (GHG) are comparatively low. Kerala has a critical eminence for Carbon Sequestration Potential due to the better geographic cover by forest. However, due to the population pressure, forest areas are under threat which leads to the higher concentration of carbon dioxide in the atmosphere. Excessive use of nitrogenous chemical fertilizers also leads to emission of green house gas (Nitrous Oxide). Similarly coconut husk retting activity in the coastal areas, discharge of sewage and domestic waste are also enhancing green house gas emission. Anaerobic conditions during wet land paddy cultivation can also result in emission of green house gas (Methane)

4. Climate Smart Agriculture

To meet the challenges of climate change in agriculture, FAO introduced the concept of Climate-Smart Agriculture at the 2010 Hague Conference on Agriculture, Food Security and Climate Change. Climate-smart agriculture does not refer to a set of practices that can be universally applied, but rather an approach that involves different elements that are embedded in specific contexts and tailored to meet local needs. Climate-smart agriculture has three main objectives:

- Sustainably increasing agricultural productivity, to support equitable increases in farm incomes, food security and development;
- Adapting and building resilience of agricultural and food security systems to climate change at multiple levels;.
- Reducing greenhouse gas emissions from agriculture
- Using Agriculture as a major tool for mitigation of Green House Gases

This does not imply that every practice applied in every location should deliver positive results for each of these four objectives. Enhancing food security while contributing to mitigate climate change and preserving the natural resource base and vital ecosystem services requires the transition to agricultural production systems that are more productive, use inputs more efficiently, have less variability and greater stability in their outputs, and are more resilient to risks, shocks and long-term climate variability. Making this shift requires considerable changes in national and local governance, legislation, policies and financial mechanisms.

4.1 Sustainable Increase in productivity and income

Climate Smart Agriculture requires a major shift in the way land, water, soil nutrients and genetic resources are managed to ensure that these resources are used more efficiently. Key strategies for ensuring that agriculture can adapt to change include:

- Sustainable soil and water management practices;
- Active promotion of biodiversity; and
- Diversification of income sources inside and outside the farms

4.1.1 Sustainable soil and water management practices

The impact of climate change will contribute to land and soil degradation and reduce the productivity of these natural resources. However, there are immense opportunities to lessen the negative impacts of climate change on land and soil resources and to optimize the potentially positive effects of climate change.

- Land resource planning is an essential entry point and process to choose the most suitable land-use systems for accommo-dating the often competing uses of land.
- Locally adapted sustainable soil and land management practices that can support climate change mitigation and adaptation.
- Increasing soil organic carbon improves nutrient and water intake by plants, which increases yields and resource efficiency of land, nutrients and water. It also reduces soil erosion and increases water retention. This combination makes the system more resilient to variability of precipitation and to extreme events.
- Agriculture sector withdraws 70 percent of total global freshwater and hence efficient use of available water will be of fundamental

importance in building resilient production systems and improving the management of climate change-induced risks.

4.1.2 Active promotion of biodiversity

Genetic resources for food and agriculture are the foundations of sustainability, resilience and adaptability in production systems. More crucially, genetic diversity is a prerequisite for adaptation and continued evolution of the species, varieties and breeds. Successful integration of biodiversity conservation into agricultural production is fundamental to maintain functioning ecosystems. Genetic diversity ensures that all kinds of life can thrive or persist under a range of environmental conditions and allows these resources to cope with pests and diseases.

There is a need to enhance conservation and sustainable use of genetic resources for food and agriculture and gather more and better information on these resources. Traditional and novel uses of genetic resources can increase the adaptability, resilience and yield of production systems and enhance their contribution to climate change mitigation. It is also widely recognized that maintaining crop diversity in production systems is crucial to avoid vulnerability and widespread crop loss resulting from particular biotic or abiotic threats.

4.1.3 Diversification of income sources

Farmer income is what ultimately counts and climate smart agriculture will be pursued by farmers only if it provides for their sustained livelihood. Many measures are being recommended as part of practicing climate smart agriculture to sustain farmer income.

- Integrated farming models like aquaculture, poultry, cattle rearing etc would be promoted to have a diversified base and make households self-sufficient in farming. It has a complimentary effect through effective recycling of wastes and crop residues and encompasses additional source of income to the farmers. It is a rich source of species diversity which helps in soil building, preserving and improving ecological conditions essential to long-term sustainability. Moreover, the emissions intensities of integrated systems are typically lower than those of specialized systems.
- Good agricultural practices, often in combination with effective input use are one of the best ways to increase small holder

productivity. It ensures farmer commitments on food production and security, food safety and quality and environmental sustainability of agriculture.

- Organic farming has potential to produce high quality food, enhance natural resource base and environment and increase income from premium price on produce apart from supporting carbon sequestration in soil. Under extreme climatic conditions such as drought which are expected to increase with climate change, organically managed farms may produce higher yields than conventionally managed ones due to improved soil properties.
- Effective use of weather information services contributes to resilience by enabling farmers to better manage the negative impacts of weather-related risks in poor seasons. Pest and disease surveillance and forecasting system is a service to the farmers, which is aimed at providing information to enable farmers of a certain area to take respective precautions with maximum economical benefit and lowest environmental hazard and to counteract eventual calamities in time.
- Greater value to farmers will arrive through assigning emphasis on post-production activities that connect the farm harvest to markets for value realisation. This will include expanding the marketing range of the farmers.
- Higher return to farmers also needs to be ensured through Agro Processing and Value Addition and marketing across the nation and world.

4.2 Adapting and building resilience to climate change

Uncertainty and risk in agricultural production need to be addressed carefully for successfully implementing climate smart agriculture.

- The natural risk in agriculture can be negotiated by appropriate weather prediction, disseminating early warning for flood and cyclones and developing contingency crop plans, climate resilient varieties, technologies etc.
- Anthropogenic risk in agriculture can be minimized through diversified cropping systems, conservation practices for carbon sequestration, water saving technologies and water harvesting development, organic farming and integrated farming systems, and forestation.

- Production risks (price risks, credit risks, technological risks and institutional risks) can be minimised by avoidance of highly risky crops, diversification of crops , weather forecast infrastructure, credit package, agriculture insurance, minimum support price system and futures markets etc.

4.3 Mitigating climate change

Climate change mitigation aims at stabilizing the greenhouse gas concentration in the atmosphere at a level that would prevent dangerous anthropogenic interference with the climate system.

- **Reducing/avoiding greenhouse gas emissions:** Improving soil management with conservation agriculture (i.e. no-tillage implemented in the context of varied crop rotations and soil protection/mulching), sustainable use of fertilizers and sustainable mechanization for crop production and improving manure management.
- **Removing greenhouse gas from the atmosphere:** This category includes policies and measures to increase carbon sequestration in standing biomass and soils. Agro-ecosystems, including forests, naturally remove carbon dioxide from the atmosphere through photosynthesis. The sequestered carbon is stored in biomass and soils, thereby acting as " Carbon sinks".

4.4 Agriculture For mitigation of green house gases

Agriculture is a major tool for mitigation of Green House Gases by its unique capacity to absorb CO_2; and release Oxygen through Photosynthesis. This could be achieved through increased cropping, by reducing rain fed areas through integrated water and river basin management, expansion of agriculture to wastelands, wetlands, degraded fallow areas and introducing urban agriculture

5. Strategies for Climate Smart Agriculture in Kerala

In Kerala, climate change has become an agenda for development planning only in the recent years. The Environment and Climate Change Department, Government of Kerala is the nodal agency for coordinating activities related to climate change in the state. There is a need to adopt a multi pronged strategy for focussed attention and implementation of Climate Smart Agriculture. It is also high time a competent authority / agency undertakes flood and drought vulnerability assessment in the major agro-climatic regions of the State. Further, Climate Smart Agriculture can succeed only if the farmers are provided with the means and measures for enhancing income and mitigating the uncertainty and risks in agriculture through credit support, crop insurance, minimum support price and market interventions.

5.1 Vulnerability assessment of agro ecological zones

Climate Change Vulnerability Assessment of each agro-climatic zones need to be undertaken for preparing vulnerability mapping and identifying vulnerable agricultural regions which are prone to climate change impacts such as flood and drought. Highly sensitive climate zones like Kuttanad region are to be declared as vulnerable agricultural zones and in these zones, climate smart agriculture protocol should be established. Vulnerable areas need to be even exempted from farming activities.

Suitability of each and every land for agriculture should be assessed and specified cropping patterns and sustainable soil management practices need to be followed. There is need for a Crop Plan strictly complying with the land use classification.

5.2 Adopting sustainable soil and water management practices

It is required to promote various soil conservation techniques and watershed management practices. Increasing soil organic matter to heighten the water retention capacity of soils is much important. Providing updated Soil Health Cards to farmers annually / biennially to farmers in all agro climatic zones in the state should be made mandatory. A mechanism for integrated management of rainwater, surface and ground water and a network of small water harvesting structures in mid land and low land areas to fully exploit the high precipitation in monsoon should be maintained. Further, we should adopt improved water management practices both on-farm and off-farm to increase water use efficiencies and reduce water pollution

5.3 Crop improvement and management

It is high time to develop new crop varieties which can survive climate extremes and are tolerant to various abiotic and biotic stresses like pests and diseases. In addition to developing improved cropping patterns with inclusion of climate smart crops, adoption of agricultural technologies such as INM, Crop Rotation etc and other farm operations should also be promoted.

Adoption of good agricultural practices, organic farming and natural farming need to be strengthened to the extent possible. Adoption of ICT / Block Chain Technology to transform the *"organic belief"* to an *"organic reality"* becomes important since mechanism currently followed is not adequate to ensure adoption of organic cultural practices. Also strict enforcement of quality control on chemical and organic pesticides being used by farmers should be enforced.

5.4 Better weather forecasting services and crop advisories

There is a need to strengthen our Weather Forecasting Services and Crop Advisories in the state. Automated Weather Stations have to be established adequately to gather ago-climatic zone based weather data and KAU should be capacitated and strengthened for facilitating weather -based and location specific crop advisories.

Maharashtra Government has installed 2,060 weather stations across the state during 2017 and 1000 more are on the way, to provide real-time weather-related information to farmers. These AWS measure wind direction, wind speed, air temperature, relative humidity and rainfall and these data are processed at the Agricultural University to issue weather and crop advisories. The advisories are shared to farmers by making them available on Mahavedh portal (Maharashtra Agriculture Weather Information Network) as well as on mobile applications. Digital kiosks in every Gram Panchayat to disburse weather related information and expert advice to farmers are also part of the programme. They also use the data to design better disaster management programmes, crop insurance schemes and to establish a weather database bank.

5.5 Information support system for pest & disease surveillance

The Information Support System for Pest & Disease Surveillance in the state needs to be strengthened. Application of Geo-informatics for real pest and disease monitoring and forecasting of possible insect or disease based timely weather information should be adopted. It is important to develop the skills and attitude of farmers towards bio-intensive pest management and to undertake research on the characteristics and spread of new pests due to variation in climate parameters.

5.6 Active Promotion of biodiversity

Genetic diversity is a pre-requisite for adaptation and continued evolution of the species, varieties and breeds. Successful integration of biodiversity conservation into agricultural production is fundamental to maintain ecosystems. Traditional and novel uses of genetic resources can increase the adaptability, resilience and yield of production systems and enhance their contribution to climate change mitigation.

5.7 Awareness creation and capacity building

Traditional wisdom should go together with modern science if we are to insulate our crops from the adverse impact of higher mean temperature. There should also be increased effort in developing cropping systems which are climate smart and nutrition rich. Climate Risk Management Centers will have to be established in every panchayat. If these steps are not taken, food and nutrition security will be impacted adversely.

Empowering farmers to better anticipate and plan for climate change is a pre requisite for practising climate smart agriculture. Awareness and assistance shall be provided to farmers for environmental modification with respect to change in climate and they should be provided with timely advise on cropping, water management and nutrient management practices.

5.8 Establishing climate smart villages

The concept of climate-smart villages is being piloted in India to help farmers cope with changing weather patterns. The project is coordinated by the CGIAR Research Programme on Climate Change, Agriculture and Food Security (CCAFS). It provided training for hundreds of farmers in technologies that help protect their crops from excessive rains and drought. In the Indian states of Bihar and Maharashtra, farmers from 40 villages have learnt how to increase the carbon content of their soil through agroforestry, effectively manage manure and apply optimum amounts of fertilizers to save money and reduce greenhouse gas emissions. Various national and international organizations are involved in the initiative, including the Indian Agriculture Ministry and Meteorological Department. The International Food Policy Research Institute (IFPRI) helps farmers prioritize adaptation and mitigation options, the International Water Management Institute (IWMI) assists farmers in dealing with water-logging through vertical drains, and the International Maize and Wheat Improvement Center (CIMMYT) provides advice on conservation agriculture and the precise use of fertilizers.

Kerala needs to adopt such a programme without delay. The focus of the climate-smart village concept is to blend local knowledge of conservation techniques with global perspectives on climate change mitigation. After selection of potential sites, a steering group of community representatives and researchers identify appropriate climate-smart options for that village. These can include climate-smart technologies, climate information services, local development and adaptation plans and supportive institutions and policies, all tailored to that community's needs. The community chooses its preferred options in a process that aims to be as participatory and inclusive as possible, encouraging participation of women and more vulnerable groups.

5.9 Market infrastructure and supply chain management

To ensure sustained farmer income while adopting climate smart agriculture, it is important to enhance the market infrastructure in the state which is currently very weak. It is required to develop decentralised village based agricultural processing centres as a support to local farmers. More agro processing and cold storage centres across agro climactic regions should be established. It is also important that the cold storage chain is adequately developed using energy efficient technologies and by adopting energy conservation measures. Value Addition and Brand Export of Kerala Products are also important in increasing farmer income.

5.10 Risk mitigation

Natural risk in agriculture can be negotiated by appropriate weather prediction, disseminating early warning for flood and cyclones and developing contingency crop plans, climate resilient varieties, technologies etc. Anthropogenic risk in agriculture can be minimized by diversified cropping systems, conservation agriculture for carbon sequestration, aerobic rice cultivation, water saving technologies, organic farming and integrated farming systems. Production risks in agriculture can be minimised by avoidance of highly risky crops, diversification of crops, weather forecast infrastructure, credit package, agriculture insurance, minimum support price system and futures markets etc.

6. Enabling Environment for Climate Smart Agriculture in Kerala

For implementing Climate Smart Agriculture in Kerala, there is a need for expanding the evidence base for determining the current and projected effects of climate change on agricultural production systems and farmers and pinpointing the key vulnerabilities. It is also required to identify and evaluate the potential climate-smart options for adapting to the expected impacts of climate change while supporting sustainable agricultural development. Farmers, pastoralists, foresters and fisher folk are the primary custodians of knowledge about their environment, agricultural ecosystems, crops, livestock, forests, fish and local climatic patterns. Efforts to adapt a climate-smart agriculture approach to a specific setting must take into account local producers' knowledge, requirements and priorities.

6.1 Support enabling policy frameworks

Supportive policies, plans and investments, and coordination in the policy-making processes and institutions responsible for agriculture, climate change, food security and land use are required to create this enabling policy framework. New policies to stimulate the adoption of climate-smart agriculture systems should focus on filling policy gaps and support the capacity development of all the stakeholders.

6.2 Strengthening state and local institutions

Enabling institutions (e.g. financial institutions, land tenure regimes, institutions regulating customary law, community-based organizations, insurance schemes, information and extension services) are essential for harnessing the evidence base to empower, enable and motivate farmers to adopt climate-smart agriculture practices. Cross-sectoral dialogues form an important part of the climate-smart agriculture methodology and it is required to enhance coordination between institutions dealing with agricultural, climate change, social protection, food security and other issues. Climate Smart Agriculture can be successfully implemented only if our local bodies conceive its philosophy and act accordingly.

6.3 Farmer friendly implementation

We will not be successful in enforcing any climate smart policies which are not farmer friendly. As with any new methods or technologies, climate-smart agriculture practices may be perceived as a risky investment. We may need to even introduce *Environment Subsidy* or *Eco Compensation* to support our farmers pursuing climate smart agriculture. Concepts like *Green Credit / Green Corpus Fund* may be established to extend the credit / insurance support to our farmers. Improved market access, which increases the returns on land and labour, is a critical factor for the adoption of new climate-smart agriculture practices. Risk Fund Concept should also be established for compensating the risk on price deficit during market glut. Innovative financing mechanisms are central to implementing climate-smart agriculture in Kerala. New climate financing mechanisms, such as Green Climate Fund, Environment Cess etc may be a way of spurring sustainable agricultural development in the State. Despite all these challenges in implementation, Climate Smart Agriculture is an *investment for the future* since there is *no future for mankind without food security*.

15

A Green Energy Future for Kerala

G. Madhusoodanan

The World Institute of Sustainable Energy with the support of WWF India concluded a one-year research study in Kerala in 2012-2013 to assess the future energy scenario for Kerala. It was an attempt to model the energy requirement (across power, transport, agriculture, industry, domestic and commercial sectors) of Kerala up to 2050 in order to assess the feasibility of meeting 100 percent of the state's energy demand with renewable sources.

The approach to the study involved four steps:

- Assessing renewable energy (RE) supply potential over the projection time frame
- Modelling of a business-as-usual (BAU) demand scenario
- Modelling of a curtailed demand scenario to assess maximum potential for demand reduction through focused measures in energy conservation (EC), energy efficiency (EE), and carrier substitution (CS). Carrier substitution is considered when the end use activity is transferred from one energy carrier to other energy carrier (for example, from transport fuel to electricity, diesel pumps to electric or solar pumps, etc)
- Matching the supply options with curtailed demand across energy carriers for electricity, heat and transport fuels.

The central finding from the study is that Kerala can meet over 95 percent of its energy demand with renewable sources by 2050. The following figure shows the final supply sources that can effect this transformation.

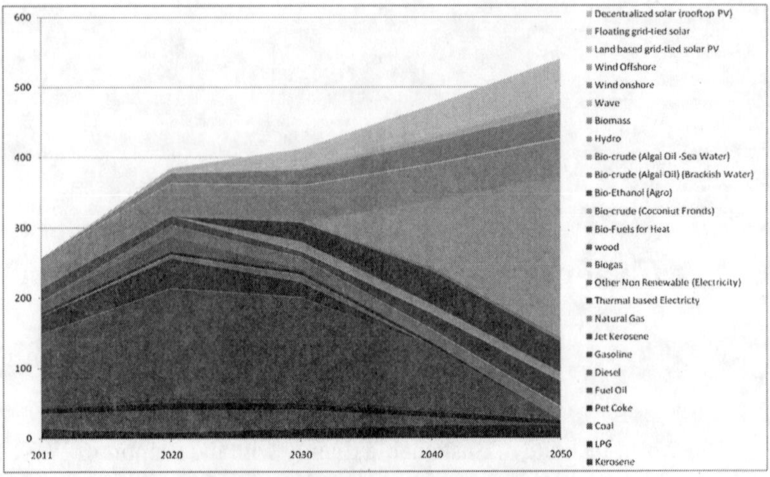

Figure-A: Supply Sources for Final Energy Scenario

The key highlights of this scenario are:

- 100 percent electricity requirements for the state can be met with RE. The main resources are onshore and offshore wind, grid-tied and decentralized/off-grid solar, large hydro, small hydro, biomass and wave energy.
- 100 percent of transport fuel requirements can be met by second and third generation bio-fuels. Considering the state of biomass to liquid and algal oil based technologies, and related technology access issues, it is assumed that these technologies will be available only by 2030. Sea based algal oil is emerging to be the main fuel supply option of the future. This assumption is consistent with past studies and the global energy report, which has assessed commercial availability of these algal based bio-fuels technologies by 2030.
- Over 70% of the total heating requirements (both for cooking and industrial processes) are met by biogas, wood and surplus bio-fuels.

The other key findings of the report are covered subsequently.

Renewable Supply potential

Table A below shows the assessed RE potential of the state

Table A: Summary of RE Potential of Kerala by 2050

Technology	Supply Potential (2050)	Remarks
Electricity	**Billion Units**	**MW (CUF)**
Grid Tied Solar PV (Wasteland)	5.99	4,273 (16%)
Grid Tied Solar PV (Grassland)	3.56	2,543 (16%)
Floating PV Panels	5.39	3,845 (16%)
Rooftop PV (Domestic)	18.33	13,079 (16%)
Roof top PV (Institutional)	25.32	18,066 (16%)
Solar Water Pumping		304 (400 hrs)
Onshore wind (Farmland) (WPD> 200)	5.98	3,103 (22%)
Onshore wind (No Farmland) (WPD>300)	0.86	447 (22%)
Onshore Wind (Plantations) (WPD>200)	8.60	4,465 (22%)
Offshore wind (WPD >250)	29.4	13,447 (25%)
Biomass gasification	0.21	37.2 (65%)
Biomass Combustion	0.62	101 (70%)
Existing Hydro (Large and Small)	11.2	1,998(65%)
Small Hydro	2.55	583 (50%)
Wave	0.37	420 (10%)
Tidal	0	0
Total	**107**	
Heat	**TJ (Max)**	
Biomass Combustion	12,206	
Biogas	18,600	
Total (Heat)	**30,806**	
Fuels	**TOE (Max)**	
Bio-Crude	296,539	
Bio-Gasoline	15,520,075	
Total	**15,816,614**	

The potential of grid-tied wind and solar energy was assessed using geographical information system (GIS). For the analysis, stringent land use and land availability criteria are considered in addition to standard technical constraints. Off-grid potential assessment is a tabulated exercise. Bioenergy assessment assesses biomass residue availability based on anticipated future trends in agricultural use. The biofuel potential assessment is based on assessed technology parameters of biomass to liquid technologies and algal oil based biofuel processing. Hydro potential assessment is based on the understanding that the only prospective new potential seems to be small hydro power. Wave energy potential estimation is based on assessment done in past studies.

Business-as-Usual Demand Projections

The BAU demand projections for Kerala indicate that the transport sector accounts for a little over 50 percent of the total energy demand of the state even in the base year 2011. This is very atypical and unexpected but validation of derived fuel demand data (from a bottom up assessment) with actual fuel use seems to indicate a good match. The BAU projections indicate an increase of about 5.5 times in total energy from 2011 to 2050. For the transport sector, the increase during this period is almost 7 times. Table B and Figure B summarize the result of the BAU demand projections:

Table B: Energy demand for the BAU Scenario

State Energy Demand (PJ) - BAU	2011	2020	2030	2040	2050
Transport	128.2	225.2	329.3	520.9	853.5
Domestic	54.4	66.7	94.5	131	169.3
Commercial	12.2	27	53	88.7	134.1
Industry	48.3	75.5	90	104.8	119.7
Agriculture	2	2.2	2.5	2.7	2.8
Public and Bulk	2.6	4.1	6.8	10.5	15.1
Total	**247.7**	**400.8**	**576.2**	**858.6**	**1294.5**

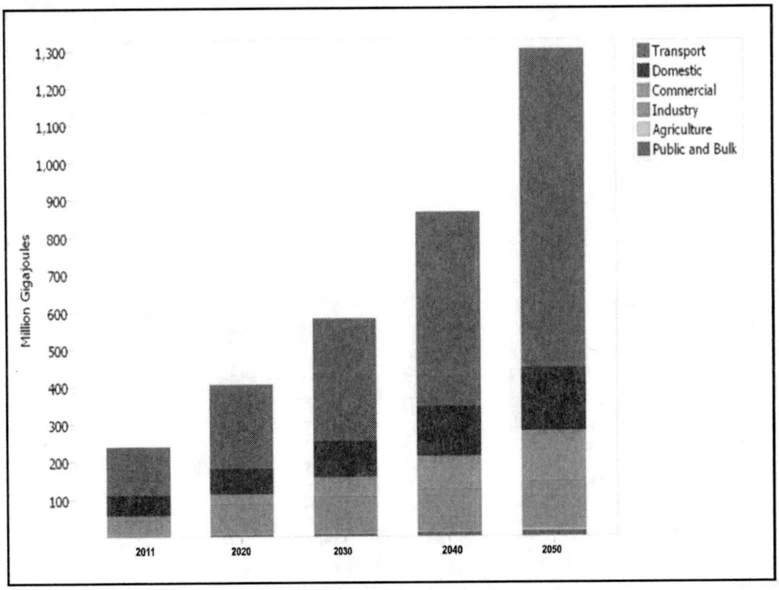

Figure B : Supply Sources for Final Energy Scenario

Curtailed demand projections

In contrast, the curtailed demand scenario indicates an increase of 2.2 times over the projection period. The curtailed demand was assessed after factoring in aggressive interventions in the BAU scenario. These interventions did not assume any reduction in economic output (or growth) but only assumed a reduction in energy intensity and better resource optimization. In economic terms, these interventions represented alternatives that effectively decoupled economic growth (GDP) from energy resource use. Table C and Figure C summarize the result of the curtailed demand scenario

Table C: Energy demand for the Curtailed Demand Scenario

Curtailed Energy Demand (PJ)	**2011**	**2020**	**2030**	**2040**	**2050**
Transport	128.2	204.8	208.8	253.3	313.7
Domestic	54.4	63.4	63.9	73.3	87.7
Commercial	12.2	20.6	36	49.2	54.1
Industry	40	81	79.3	73.9	63
Agriculture	2	2	1.8	1.8	1.8
Public and Bulk	2.6	3	5.1	8	11.6
Total	**239.4**	**374.8**	**395.1**	**459.5**	**531.8**

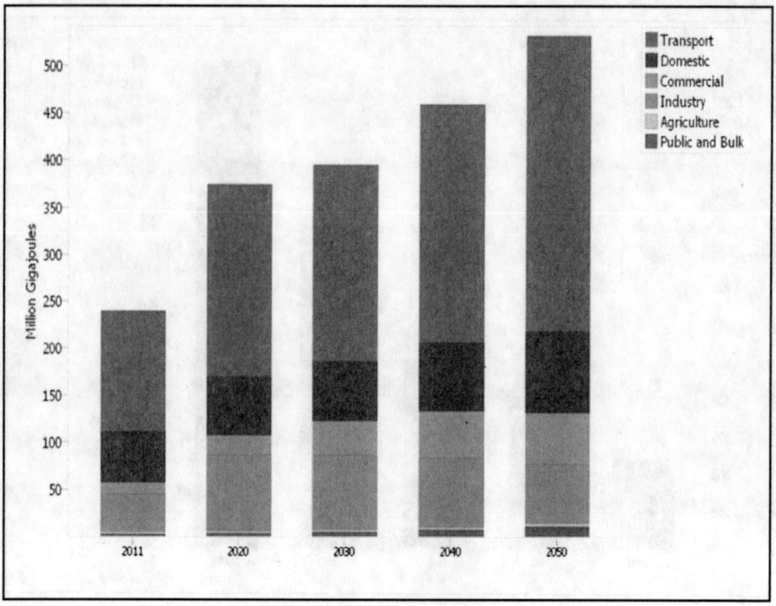

Figure C : Sectoral Energy Demand Projections (Curtailed Demand Scenario)

On a broad level, the project findings indicate that while the existing pattern of growth (BAU growth) would lead to overdependence on fossil fuels, aggressive interventions in energy efficiency, energy conservation and carrier substitution can curtail demand significantly. More importantly, the curtailed demand scenario has the potential to drastically reduce the state's dependence of fossil fuels to achieve a near 100 percent RE supply by 2050.

16

Sustainable Agriculture : Past Lapses and the Way Ahead

Mary George

Kerala agriculture is characterised by low scale and low productivity. Around 96 percent of operational holdings are below one hectare (Economic Review 2018, Appendix Table 2.30). The problem is exacerbated by weather and market risks. Around 80 percent of the area under cultivation does not have access to irrigation. Ramesh Chand et.al. (2017)[1] highlights the finding of the National Sample Survey on situation Assessment survey of Agriculture households (NSS-SAS), that 13.9% farm households in India experienced negative return from crop production during 2012-13. Non-farm income comprised 40% of the income of farm households, but access to non-farm sources of income is highly skewed as about 40 percent of farm households reported zero income from such sources. In the case of Kerala, non-farm source of income in the rural sector is significant with its absence.

Kerala, with a total geographical area of 38.8 lakh hectares, suffers seriously from the hazards of soil erosion due to factors like high but erratic rainfall, undulated topography, indiscriminate deforestation, encroachment by land lobby, unchecked and illegal quarrying and sand and other mineral mining. Planning Board (1984)[2] reported that watershed method was followed as the method of planning and execution of soil conservation measures. Out of the total geographical area 38.9% or 15 lakh hectares were identified as erosion prone. Only 75,000 hectares or 1.9 percent of the erosion prone land underwent soil erosion treatment until mid 1980's. (Report of the High Level Committee on Land and Water Resources, State planning Board, 1984)

Food Security at Risk

The land utilization statistics of the state underlines that 24.84 lakh hectares or 66.5 percent of the geographical area is under various crop cultivation. Cropping pattern underwent radical shift between 1952-53 and 2016-17. Area under food crops which was 7,88,190 hectares in 1952-53, increased to an all time high of 9,26,350 hectares in 1970-71. Around this time area expansion reached a peak and there after declined to 8,01,699 hectares by 1980-81. It further declined to 5,59,450 hectares in 1990-91 and to 1,71,398 hectares by 2016-17. At present only 6.6 percent of the total cropped area is under paddy cultivation. Thus Kerala's food security is highly at risk. According to United Nations Committee on World Food Security, food security is realized when all people at all times have physical, social and economic access to sufficient, safe and nutritious food to meet their dietary needs and food preferences for an active and healthy life. Nutritional dimension is all the more important. To reach this goal, U.N recommends to promote sustainable agriculture as a sustainable development goal (SDG) by 2030. Food security according to UN can be reached only by doubling of food output by the stipulated period by developing countries like India. In the Indian Federal set up Kerala is going to be the worst victim of food insecurity. Food security Act 2013, promises to give money worth of secure food, if the centre fails to safeguard food security provisions of states through Public Distribution System. When the centre fails to procure food grains from within the country or outside how can states find sources of getting it? Hence Food Security Act does not guarantee right to food of the citizen. Similarly, for fruits and vegetables, Kerala heavily depends on neighbouring states. Excessive and unscientific use of crop protection chemicals by greedy and unscrupulous farmers results in high level contamination of food grains, fruits and vegetables which enter the food chain of Kerala. While India is identified as a country with high incidence of cancer, Kerala has 3 times higher cancer incidence than the national average, mainly because of the pesticide contaminated food intake.

Plantation Crops Sector Under Stress

Major plantation crops in India are rubber, coffee, tea and cardamom. Government of India from 1940's and 1950's encouraged Kerala to concentrate in the cultivation of plantation crops as the topography and climatic conditions of the state are suitable for that. These crops are also of great economic importance and play a vital role in bringing foreign exchange through exports and generate employment to millions. These plantation crops together occupy 7.05 lakh hectares, accounting for 34.6 per cent of the net cropped area in the state (Appendix 2.4) Economic

Review 2018). Kerala's share in the national production of rubber is 78 per cent, cardamom 88 per cent, coffee 21 percent and tea 4.7 percent in the year 2017-18 (Appendix 2.11 economic Review 2018). In the recent years, ailing plantation industry is yet another cause of worry. Cost of production in the plantation sector has been rising rapidly, thanks not to any modernisation programme, but to increasing labour and welfare costs which together makes up 65% of the total cost of production. Plantation industry in Kerala accounts for 33 percent of agricultural GSDP. It directly employs 3.5 lakh workers. Out of the total plantation crop production of the country, 45 percent is the contribution of Kerala. Therefore, technology up-gradation and modernisation, value addition etc are very essential for the sustainability of the industry. The 2019-20, state budget has mentioned about the creation of a CIAL model corporation for value addition of rubber. Similar other projects are to be launched for the sustainability of other crops. Branding of coffee announced in the state budget is a consolation for the coffee sector. Many new innovative start-ups are essential for the growth of farm income and non-farm sources of income and employment in the state.

Contribution of Primary Sector to GSDP

Agriculture underwent drastic changes over the years. Shift in cropping pattern, slow growth, stagnation, recovery etc. have been the crucial features exhibited by agriculture. Shares of each sector to Gross State Domestic Product will reveal the relative performance of each sector where agriculture is the major component of the primary sector

Table 16.1 : Sectoral Distribution of Gross State Value Added (at constant prices 2011-12) (Rs. in Lakh)

Area Production and Productivity, Principal crops:

Item	2011-12	2012-13	2013-14	2014-15	2015-16	2016-17	2017-18
						(P)	(Q)
Primary	15.19	14.4	13.29	13.35	11.34	10.94	10.85
Sector							
Secondary	27.3	26.53	26.3	25.89	27.21	27.25	27.40
Sector							
Teritary	57.45	59.06	60.38	60.76	61.64	61.81	61.75
Sector							
Total GSVA	100.00	100.00	100.00	100.00	100.00	100.00	100.00
At Basic Pries							
Total GSVA							
At Basic							
Pri.	**33629311**	**35635473**	**37165147**	**38586962**	**40648007**	**42545868**	**45074499**

When Table 16.1 is examined, the relative share of primary sector is found declining from 15.19 in 2011-12 to 10.85 in 2017-18.

Area Production and Productivity of Principal Crops (1980-81 to 2016-17)

Kerala's unique and diverse agro-climatic conditions permit it to go for a variety of crops. Earlier food crops occupied a premier position in cropping pattern. But now food crops comprising rice, tapioca and pulses accounts for only 10.12 percent of the total cropped area, while cash crops-cashew, rubber, pepper, coconut, cardamom, tea and coffee-occupy 61.6 percent of the cropped area (Eco. Review, 2018). When 15 major crops are taken into account they cover more than 80 percent of the cropped area of the state.

Table 16.2 : Compound Growth Rate of Area, Production and Productivity of Principal Crops (1980-81 to 2016-17)

Crops	Area	Production	Productivity
Rice	-4.16	-2.57	1.66
All Pulses	-6.86	-6.46	0.43
Turmeric	-1.007	-0.042	0.975
Rubber	2.68	5.97	3.20
Tea	0.064	0.45	0.39
Coconut	0.62	2.21	1.58
Areconut (1979-2017	2.62	14.07	11.16
Pepper	1.61	-0.003	-1.58
Ginger	-2.89	-0.39	2.58
Tapioca	-3.99	-1.67	2.43
Benana	4.497	2.913	-1.52
Other Plantain	1.49	4.021	2.397
Cashew nut	-3.56	-0.97	1.112
Coffee	0.644	2.702	2.045

Source: Eco Review, Various years

When Table 16.2 is analysed, a single crop which lost largest area between 1980-81 and 2017 -18 is rice in which case compound growth rate of area decline was -4.16. Area under rice declined from 8,01,700 hectares in 1980-81 to 1,71,400 hectares 2016-17. When tapioca lost its area by -3.99 percent, cashew lost it by -3.56 per cent. Food security is endangered when rice and tapioca lost area under cultivations on a large scale. Similarly, nutritional security is at risk when area under all pulses declined by -6.86 percent. Pulses are protein rich and the major source of protein for marginalised classes. Production of pulses is encouraged by the central government under "rainbow revolution" to overcome India's highest level of malnutrition among world nations. Major crops which underwent area expansion are banana (4.49%), rubber (2.68%), arecanut (2.62%), other plantation (1.49%) and coffee (0.64%). Inspite of the decline in area under cultivation, because of the improvement in productivity, the decline in production was less than the percentage decline in area between the two periods for those crops which largely lost area under cultivation.

Why massive shift has taken place from food crops to non food crops. Various reasons are cited for this. Joshi et al (2003) found that when a hectare of rice cultivation in one season requires 105 labour days, that required for wheat is 44 labour days. Joshi P.K and Ashok Gulati (2003)[3], "From Plate to Plough; Agricultural Diversification in India" (IFPRI Conference). Agriculture wages in Kerala are akin to wages in the west. But in the West man days are replaced by Machine hours largely. Scarcity of agricultural labour coupled with high cost of labour and other inputs made rice cultivation a loss making enterprise and area under rice was mostly shifted to other crops or kept fallow. Rapid urbanization of Kerala added fuel to fire. Land prices sky-rocketed and it became a better asset than even gold. Land market was booming and people found it better to sell land and keep the proceeds in the bank.

Input use and Productivity

When labour cost has been exorbitant, prices of other inputs have also been very high.

Table 16.3 : Price and Consumption Trend of Chemical Fertilizers (2015 to 2018)

(Prices Rs./ Metric Ton) **(Consumption in Metric Ton)**

Item/Year	Prices			Consumption		
	2015-16	2016-17	2017-18	2015-16	2016-17	2017-18
Urea	5413	5684	5900	1,30,547	78580	138213
Amonium sulphate	13,000	13,000	13140.18			
Phosphorus	7200	7680	7327.63	48879	43192	43295
Potash	16000	12167	11868.35	81004	58644	60472

Source: Economic Review 2018 (Appendix 2.21; 2.22)
**Urea and Ammonium sulphate represent nitrogenous fertilizer and hence consumption of both comes under Nitrogenous fertilizer.*

From the table it may be understood that high price of chemical fertilizers results in lower consumption. Before and after globalization price increase of fertilizer was absorbed by increased subsidy. But, the

NDA Government's declared subsidy policy has been one of gradual reduction and final elimination of subsidy. Therefore price rise of any fertilizer, other things being equal, results in reduced consumption. Chemical fertilizer has a catalyst's role in raising productivity, if it is applied as per recommended doze, accompanied by irrigation.

Irrigation

Kerala agriculture which depends on the monsoon is a gamble. Normal rainfall is declining. Irrigation potential created (IPC) is high as there are forty four rivers and 36 dams in the state. But irrigation potential utilised (IPU) is very low.

Table - 16.4 : Net Area Irrigated (Source Wise) (in hectare)

No.	Source	2015-16	2016-17	2017-18
1.	Govt. Canals	88817	63403	76761
2.	Private Canals	774	932	612
3.	Tanks	48459	49657	49773
4.	Wells	133529	122478	123115
5.	Other Sources	142254	141291	141745
6.	Total	4,13,833	3,77,761	3,92,006
7.	Gross Irrigated Area	4,83,648	4,97,386	5,39,874
8.	Net Area Irrigated to net Area shown (%)	20.45	18.74	19.21
9.	Gross Irrigated area to gross cropped area (%)	18.40	19.25	20.92
10.	Irrigated area under paddy to total irrigated area	31.12	26.35	26.93

Source: Economic Review: 2018
Appendix Table 2.69

Net irrigated area is around 20 percent when 44 percent is the national average. At the all India level, 93.6 percent of the rice cultivation is irrigated while in Kerala it is only 26.93 percent in 2017-18. Low level of irrigation, low level use of chemical fertilizers etc. mean low level of production investment which throws light on the distress situation of Kerala Agriculture. If production investments were high, price disadvantages could be overcome with productivity advantages. But, because of the low input use in agriculture, most of the crops enjoy lower productivity than the national average. Swaminathan (1971)[4] asserted that the synergistic package of practices comprising the use of seeds of high yielding hybrid varieties with adequate and timely supply of plant nutrients through fertiliser and water followed by adoption of suitable agronomic technology and prophylactic measures as well as post-harvest technology of storage, processing etc.are the pivot of successful agriculture. The synergy or the mechanism that makes the whole something very much more than the sum of the parts is a potent tool in the nature of getting the large effect from small resources. The release of synergetic-interactions provides the best possible means of enhancing the cost benefit ratio of developmental strategies (M.S.Swaminathan; 1971; synergestic effect of co-ordinated use of Fertilizers and other inputs; Fertilizer News, January 1971 P.45).

Declining Institutional Credit

Credit is the mobilizer of agricultural resources and life-line of production. But there are many a ways of diversion of agricultural credit to non-agricultural zones. Yet another challenge is the increase in investment credit in relation to production credit. When production credits are meant to meet input costs and other related expenses, investment credits are not crop loans. Hence, farmers approach non-institutional lenders who charge exorbitant rates on loans. Once a farmer falls in to the clutches of such money lenders, there is no escape from them. The pattern of agricultural credit would be clear from the following table.

Table - 16.5 : Break up of Production & Investment credit (Rs. Crores)

Year/Item	Production credit	Investment credit	Column 3;Column 5	Total Credit
2013-14	36872	6019	14.03	42891
2014-15	41685	7395	15.07	49080
2015-16	41854	10519	20.08	52373
2016-17	40410	13860	25.54	54270
2017-18	48243	18847	28.09	67089

Source: Economic Review; 2018; Appendix 2.23

Table elucidates the increase in the investment credit at the expense of production loan or crop loan.

Table - 16.6 : Yearly Average Index Number of Prices Received and Prices Paid by Farmers (in Rs.)

Year	Index of Prices Received by farmers	Index of Farm Cultivation cost	Index of Domestic Expenditure	Index of prices paid by farmer	Party Index as column 2 to 5
2007	2746.00	9135.00	3061.00	5288.00	52.00
2008	3167.00	10036.00	3394.00	5837.00	54.00
2009	3083.00	11468.00	3640.00	6460.00	48.00

(Contd...)

2010	5117.00	14959.00	4398.00	8109.00	63.00
2011	4743.00	17447.00	4715.00	9070.00	52.00
2012	6008.00	20422.50	5376.58	10478.42	57.08
2013	8272.25	22407.33	5879.50	11447.67	72.08
2014	7706.08	20422.50	5376.58	10478.42	57.08
2015	7730.75	25773.42	6493.17	12936.25	59.83
2016	8862.25	27106.75	6844.08	13620.00	65.08
2017	9019.50	28122.17	6984.83	14015.00	64.17
2018	5117.00	14959.00	4398.00	8109.00	63.00
2010	4743.00	17447.00	4715.00	9070.00	52.00

* Average of monthly indices up to June 2018

Source: Economic Review Various Issues, State planning Board 2018

Farm Distress and its Determinants

Agrarian distress in India, particularly in Kerala may be associated with low agricultural prices and consequently poor farm incomes. Supply side factors like low credit availability, low input use and low productivity are equally important. Prices play a key role in affecting the incomes of farmers. In the last two years, inflation in agriculture is much lower than, overall inflation. Rangarajan et al (2019)[5] observed that the implicit price deflator for Gross Value Added (G VA) in agriculture was 1.1% while it was 3.2% for total GVA in 2017-18. The advance estimate for 2018-19 show that the implicit deflator for GVA in agriculture is 0%, and 4.8 % for total GVA. In fact, agriculture GVA growth was at 3.8% for both nominal prices and constant prices in 2018-19, giving the price deflator of 0%. (Rangarajan and S. Mahendra Dev 2018; Removing the roots of Farmers' Distress, The Hindu Jan 28/2019). Market prices for general agricultural commodities have been lower than their minimum support prices. Those who get the benefit of the minimum support prices are those farmers with marketable surplus. Since 96% of the Kerala farmers have less than one hectare, farmers with marketable surplus are very small in number. Because of productivity advantages, when output increases beyond the market demand at a price remunerative to farmers, market price declines. In the absence of an effective price support policy, farmers meet with this type of low returns year after year. To overcome this, there should be sustained increase in output on the one hand and sustained increase in price support on the other.

Rural Kerala in Debt

NSSO's Debt Deposit Survey of rural and urban households (2012), placed Kerala in high debt category states. In rural Kerala, the average debt of a household is Rs. 1,47,402. When other charges are added to it total amount to be repaid becomes Rs. 2,97,752. Debt burden of rural Kerala varies between rural agricultural households and rural non-agricultural households. Out of the rural agricultural households 68.07% is under debt. Average debt burden is Rs. 4,41,589 and it becomes Rs. 6,48734 when the total amount to be repaid is taken into account. It should be noted that 49.10% rural non agricultural households are in debt. Average debt is Rs. 1,41,029. Debt along with interest rate and other dues make it Rs. 2,87,212. These findings are corroborated by the findings of NABARD in its All India Rural Financial Inclusion Survey 2016-17. Kerala, with 56% rural households under indebted condition comes 6th among the states of India, topped by Telangana with 79% households in debt.

Loan Waiver - A Poor Solution

The amount of loans taken from institutional and non-institutional sources for agriculture exceeds agricultural requirements. This indicates that a part of the crop loans is spent on non-agricultural purposes. A more worrisome finding of NSS surveys on Investment and Debt-(NSS-I &D) is that the loans taken by cultivators from non-institutional sources which involve high interest rate is rising faster than from institutional sources. Ramesh Chand et al (2017)[6] found that much of the growth in household demand in rural India has been debt-ridden and not supported by growth in income (Rameshchand and S.K. Srivastava 2017, "Think Beyond Loan Waivers", The Hindu, July 20, 2017). Very often good number of eligible farmers are excluded from agricultural loans and also subsidy/loan waiver benefits etc. CAG, reports many startling findings with regard to identification of eligible farmers for loans, Kissan Card beneficiaries, loan waiver beneficiaries etc. According to CAG, 13.46% of the accounts which were actually eligible for the benefits under the scheme were not considered by the lending institutions while preparing the list of eligible farmers. Out of those who enjoyed the privilege of debt waiver, 8.5% were neither farmer nor eligible for any benefits, similarly, 34.28% of the beneficiaries were not issued debt relief certificates which would have entitled them to fresh loans. Further, NSS-SAS study found that about 39% of farmers borrowed only from non-institutional sources.

Divided Opinion on Loan Waivers

Economists and policy makers are divided on loan waiver policy. Former RBI governor, Raghuram Rajan had negative views on debt waiver.

Reasons he cited are: First, loan waivers have "reputational consequences", that is, they adversely affect the repayment discipline of farmers, leading to a rise in defaults in future. Second, earlier debt waiver schemes have not led to increase in investment or productivity in agriculture. Third, after the implantation of debt waiver schemes a farmer's access to formal sector lenders declines, leading to a rise in their dependence on informal sector lenders. Based on the lessons learnt from the two major loan waiver policies of Independent India, one in 1990 and the second in 2008 Ramakumar makes certain valid observations. In September 2018, agricultural non-performing assets (about 8%) were far lower than in industry (about 21%). Further agricultural NPA's were on a continuous decline between 2001 and 2008. In 2018 RBI sources reported that around 80 percent of the NPA's were accounted for by the industrial sector.

In the recent years when agricultural growth rate is very low or negative, farmers are in great distress. Farmers protests and agitations are on the rise. This pinpoints the gravity of the distress. Some of the state governments, have already implemented loan waiver policies. Eleven states have announced schemes for loan waivers. In Kerala, distress is mounting up; In Wayanad and Idukki and the other districts badly hit by floods are in great distress and farmer suicides are becoming frequent.

Measures to Promote Sustainable Agriculture

Sustainable solution to indebtedness and agrarian distress are to raise agricultural income and to enhance access to non-farm sources of income. The low scale of farms necessitates that some cultivators move from agriculture to non-farm jobs. Improved technology, expansion of irrigation, crop diversification towards high value crops etc are measures to improve farm income. Important steps that may be taken by the government to redress the distress of farmers are:

(a) Reasonably high minimum support price

Announcing higher minimum support price is not enough. Because, in that case MSP benefit will reach only large farmers with marketable surplus. As Punjab did in 2017, government should purchase all food grains at the MSP. This facility should be extended to all crops, whether in distress or not. This will enable farmers to have some income that will make agriculture meaningful. Unless there is a procurement guarantee and the farmer can count on the government to purchase his produce, the MSP has little meaning on the ground.

(b) E-NAM or electronic national agricultural marketing

E-NAM, is expected to find solutions to all problems afflicting agriculture and is supposed to be fully operative all over India from 2018. But so far very few steps are taken by the centre as well as by the states to prepare the needed single electronic platform for sale across the country. The central government along with state government has to create congenial atmosphere and infrastructure for the smooth functioning of the E-NAM Platform. Essential facilities include:-

- Facilities to assure the quality standards as per national and international standard stipulation.
- Those who directly deal with the market-producer (sellers) and ultimate buyers, must get remunerative prices, grading, direct payment of sale prices, transporting product to the buyers destination etc. Central government plans to introduce E-NAM in selected 585 'mandis'. For this, needed infrastructure must be prepared and provided by the state government. State government has to ensure that:-
- With a single licence, trading from any part of the state or country must be possible.
- Transaction fee remitted in any one point must be sufficient to trade in any part of the country.
- Infrastructure demands huge investments. Vegetables and fruits need refrigerated godowns, grading and assaying (quality testing). Grading, branding etc should follow international stipulations.

E-NAM is partially implemented in Karnataka and compared to other farmers they are getting 16% higher price for agricultural commodities. If E-NAM is made fool proof, farmers' income appreciation would be higher as middle man would disappear.

(c) Improve water use-efficiency

Per capita availability of water is decreasing in India. In 1999, the average per capita availability of water was around 2,200 cubic meters (cm) which has fallen to about 1829 cm. in 2018. It may further go down as population increases. It may fall down to 1340 cm and 1140cm respectively in the years 2025 and 2050. Though Kerala has achieved replacement rate of growth of population, because of the influx of migrant population, water need may rise. If per capita availability of water falls below 1700 cm, then the state will be considered "water stressed" and if it falls below 1000 cm then "water scarce". Water use efficiency is very low in India and it is worse in Kerala. Irrigation potential created in Kerala is much larger than the potential utilized. More crop from every drop should be the norm of water use and every source of water harvesting must be tapped. Polluted water courses should be given re-birth through purification measures.

To make agriculture sustainable in the state, the following steps should be taken urgently.

- Value addition of agricultural products and creation of non-farm employment opportunities will raise farmers' income on the one hand and more employment in the rural sides.
- Mixed farming. Instead of growing mono crop, simultaneously grow a few crops like rice, fish and poultry. Income loss from one source would be made up by gain from other. Loss minimisation and income maximisation is sought from mixed farming.
- Creation of producer companies for value addition like the one announced in the budget 2019-20, viz, CIAL model rubber value addition company
- Coconut based products. Starting from virgin coconut oil, more than 40 value added products may be turned out from this single resource many of which would capture world market. 'Neera' is in crisis simply because of wrong policy pursued by the presentgovt.
- Land Banks: Those who leave their land fallow may be tempted to hand over the land to the land bank, without loosing ownership rights. Lease- land cultivation should be promoted. Rent collected, after deducting the service cost may be transferred to land owner which becomes his revenue.
- Proper soil test and soil health card should be introduced. After knowing the nutritional status of each region, suitable organic and inorganic manures may be applied as per the recommended doze for each crop. Chemical fertilizers are like Allopathy medicines and application of recommended doze, as prescribed, would do wonders.
- Compulsory certification of organic agro-products. Every organic producer should get the opportunity to produce his own brand. To speed up the process of organic certification, government

through Agriculture officer should extend, technical and monetary help.

Maintenance of ecological balance necessitates maintenance of wetlands. Palakkad district which accounts for 38 percent of the remaining paddy fields in the state was the minimum affected by the floods of 2018. The change in land use pattern and water management has resulted in a sharp decline in the area under paddy from 35 percent of the cropped area in the 1960 to 6.6% in 2018. If agriculture is any goal, then paddy land conversion should be banned. Climate change and water scarcity that Kerala faces today is largely linked to paddy land conversion. The Kerala Conservation of Paddy Land and Wetland Act 2008 should be rigorously implemented with retrospective effect. Gadgill Report of 2011 should be implemented. Post Disaster Needs Assessment (PDNA) report, prepared by a team of 76 international experts in collaboration with the state government estimated flood loss as 31,000 crore. Finance minister calculated the fund needed as one lakh crore, which seems to be more accurate when we consider the period needed to regain the loss in terms of area, crops, houses and lively-hoods lost.

Crop diversification is a double edged sword. It improved farm income on the one hand and ecological balance on the other. On the other hand, Kerala instead, moves from crop diversification to crop concentration. Crop diversification is socially and economically beneficial, when complimented by infrastructure facilities, finance and technical support especially for local micro-enterprises, which are engaged in processing, storing, grading and packaging.

- Kudumbashree (women self & help groups in Kerala), has successfully adopted crop diversification. Mary George (2007)[7] found 147 varieties of Kudumbashree micro-enterprises. Among them most sustainable were crop diversification lease-land enterprises. (Mary george 2007; Technological sustainability of women enterprises under kudumbashree programme of Kerala; Report submitted to DSIR, Ministry of Science and Technology, Govt of India)
- Strict sanitary and phyto-sanitary standards are to be adopted to make agricultural products and agro-based value added products internationally competitive.
- Mahatma Gandhi National Rural Employment Programme Mission should be entrusted with the duty of water harvesting, recycling of drainage water, cleaning up of water courses etc.

In order to safe guard farmers' crops from wild animals protective mechanisms may be created with MGNREGP work force. Construction of concrete trench between forest and crop land to prevent the entry of wild elephants is one such measure. Similar, suitable precautionary constructions may be made to prevent the encroachment of wild-boar, monkey etc.

- Agricultural budget must attract at least 33% of the plan fund- because sustainability of any country/ state, ultimately depends on sustainable food security, safe environment etc.
- Mechanization of agriculture and making agriculture a white collar job are essential to attract youth to agriculture. A corpus Fund for agriculture should be created. Whenever salary revision is undertaken, for govt and quasi-govt employees, deserving sum should be directed to the corpus fund. This fund may be used to give bonus, festival allowance, crop insurance etc to the farming sector.
- Karshaka Labour Bank of Vadakkancherry model may be formed. In the process of mechanization of agriculture agricultural workers may be given training for the proper use of inputs including pesticides and fertilizers.
- Information technology platform may be widely used to give fast information on climate change, demand/supply/price variations within the country and across countries.

More than 80 percent of the farmers in Kerala grow their corp with rain water. A new Rain Fed Agricultural Atlas released in 2018 reveals the 'sheer negligence' towards rain-fed areas. According to Ashok Dalwai, CEO of the National Rain-fed Area Authority and the Chairman of Committee on Doubling of Farmers Income, "Farmers in rain-fed areas are receiving 40% less of their income from agriculture in comparison to those in irrigated area". This observation calls for the attention of the Government to invest more in research and development in rain-fed agriculture.

Conclusion

Rural Kerala is in debt. Farmers are far more deeply indebted. National Sample Survey Organization Survey, National Bank for Agriculture and Rural Development Survey etc sheds light to the grave situation of Kerala agriculture. The most backward tribal districts of Kerala-Idukki and Wayanad-are the worst victims of the 2018 flood. Eight thousand farmers in Wayanad and fifteen thousand in Idukki are under the threat of eviction owing to non-repayment of bank/co-operative loans. Most of them are in the clutches of money lenders. Sino-Indian team of economists Shengen Fan,Ashok Gulati and S Thorat, in their study found that in India, investment is very low in health, education, and rural roads. Though public investment was very high during the green revolution period, later it was drastically cut. This has adversely affected the farm sector growth. As a result, as Pulapre (2019)[8] puts it "the price of food has historically been high for Indians at the bottom of the income distribution. This has held back industrialisation and the desirable shifting of population away from farming to other activities" (The Hindu, Feb.8, 2019). Flood devastated Kerala agriculture and its dependents, looks for alternatives. Kerala has always been food deficient. Plantation sector has also been failing after the removal of Quantitative Restriction on imports (QRs) of 824 agricultural commodities after signing the WTO Treaty in 1995. A new, scientific look at agriculture, with a shift to high value crops, without discarding the traditional varieties, without eroding ecological balance, is what Kerala is looking for.

References

1. Rameshchand et. al (2017): National Sample survey on situation Assessment Survey of Agriculture households (NSS-SAS).
2. Planning Board (1984): Report of the High Level Committee on Land and Water Resources, State planning Board.
3. Joshi P.K and Ashok Gulati (2003): "From Plate to Plough", Agricultural Diversification in India (IFPRI Conference).
4. Swaminathan (1971): M.S Swaminathan; 1971; synergistic effect of co-ordinated use of Fertilizers and other inputs; *Fertilizer News*, January 1971 P.45.
5. Rangarajan et al (2019): (Rangarajan and S. Mahendra Dev 2018; Removing the Roots of Earners Distress, *The Hindu*, Jan 28/2019.
6. Rameshchand et al (2017) - (Rameshchand and S.K. Srivastava 2017, "Think Beyond Loan Waivers", *The Hindu*, July 20, 2017).
7. Dr. Mary George (2007) - (Mary George 2007; Technological sustainability of women enterprises under kudumbashree programm of Kerala. (*Report submitted to DSIR, Minister of science and Technology,* Govt of India)
8. Pulapre Balakrishnan, (2019); *The Hindu*, Feb.8, 2019.

17

Rebuilding Kerala: Need to Integrate Animal and Human Health

Gayathri Balagopal

Introduction

Among all pathogens that affect human population, nearly 60 percent are transmitted from animals. The World Health Organisation argues that the interconnectedness between humans, animals and the environment could be a source of diseases such as leptospirosis, scrub typhus, kyasanur forest disease, crimean-congo haemorrhagic fever, rabies, avian influenza, brucellosis etc. that spreads from animals to humans through direct contact, water and air, which impacts public health and livelihood. As a result of changing agricultural practices, human settlement near or in forests, improper management of water bodies and inadequate waste management practices, the spread of vector-borne infections like dengue and chikungunya are on the rise. Anthropogenic and natural changes to the ecosystem can result in increased interaction between reservoirs, vectors, pathogens and human hosts, with vectors selecting anthrophyllic hosts over zoophyllic hosts (Morse 1995). According to Dubbal *et al* (2014), there is a public health and economic risk associated with close relationship with humans and a wide range of animals like pet animals, companion animals, wild animals and synanthropic animals in developing countries. In addition, food-borne infections are a source of concern, with pesticide residues in plant-based food, antibiotic residue in animal-based food and bacterial contamination in water, plant- and animal-based food.

Factors like rainfall and temperature are known to play a role in creating favourable conditions for breeding of vectors like mosquitoes. In a study on five states, namely, Kerala, Haryana, Punjab, Gujarat and Rajasthan, it was found that Kerala had the highest number of dengue cases probably due to having more breeding spaces for mosquitoes, suitable temperature ranges (23.5–30°C) and short incubation period particularly during rainy seasons(Mutheneni *et al* 2017). In addition, the higher transmission of dengue during monsoon and post monsoon period was influenced by rainfall and ambient temperature in these states (Chakravarti *et al* 2012). Besides seasonality, certain practices like collection of water in containers within the house or compound was found to a statistically modifiable predictor of the occurrence of chikungunya in a house, according to a study conducted in rural areas of Thiruvananthapuram district in Kerala (Anish *et al* 2011).

Deforestation is often accompanied by introduction of agricultural activities and plantation industries, which bring in human settlements near the edge of the forest. Further, people are employed in economic activities like gathering non wood forest products and transporting logs from the forest. Literature highlights the role played by policies such as the conversion of forest land to plantations in the late nineteenth century, conversion of forest land to grow food crops in the 1940s, creation of new settlements in forests in the 1950s and 1960s and development of irrigation and transport systems in forested areas in the twentieth century, in deforestation in Kerala (George and Chhattopadhyay 2001). These changes to the eco system exposes people to animal reservoirs like fox and bats, ticks that were zoophyllic and vectors that flourish in dense vegetation. Thus the opportunities for transmission of pathogens increase (Patz *et al* 2000). In Kerala, Wayanad Wildlife Sanctuary has experienced deforestation due to conversion of forests to plantations between 1920 and 1973, which has resulted in a loss of forest cover of nearly 23.1 percent (Satish *et al* 2014). The loss of forest cover signifies that animal population in the forest may have been affected, with a reduction in predator population like leopards and tigers and increase in their prey like deer. This could be a factor behind outbreak of Kyasanur Forest Disease (KFD) and Lyme disease in Wayanad. Both these diseases are spread through the bite of ticks. With the encroachment of human population into or near forested areas, they serve as new hosts for ticks, which then transmit these diseases. In 2013, there was an outbreak of Lyme disease in Wayanad among workers in a coffee plantation and villagers did highlight the growing menace of deer population near their habitation (The Times of India 2013). Kerala had to deal with an outbreak of a highly fatal zoonotic disease, Nipah virus transmitted from fruit bats to humans, which is related to ecological imbalances like loss of natural habitat of fruit bats.

According to the Inter-State Migrant Survey in Kerala, there were 25 lakh migrant workers in the state (India Today 2017). Inter-state migration of labourers to Kerala from states where diseases like Kala Azar, cholera are known to be endemic could also be risk factors for the emergence of these diseases (Sukumaran and Pradeep Kumar 2015). International migration and travel can also serve as a conduit for carrying highly infectious diseases. Recently, a man who had worked in a slaughter house in United Arab Emirates and returned to Kerala was admitted to hospital with Crimean-Congo Haemorrhagic Fever (The New Indian Express 2018).

The growing problem of roadside dumping of municipal waste is a reality in most parts of Kerala. With most local bodies adopting the no-bin approach to waste, lack of centralised waste management systems due to citizen protests and with households not being aware of managing food wastes in their homes, a large amount of garbage is dumped on the streets. The Swachhta Status Report highlights that 87.7 percent of rural households and 75.7 percent of urban households in Kerala did not have garbage disposal arrangement (NSSO 2016). While many households in the state will have arrangements within their premises to dispose of food waste, those households that lack such arrangements or lack awareness tend to dispose their garbage on the roadside. This has contributed in the growth of stray dog and cattle population and in attracting rodents. Dogs, cattle and rodents can transmit leptospirosis to human population.

After the 2018 floods in Kerala, there was a spurt in zoonoses like leptospirosis and scrub typhus. Given that Kerala has one of the highest human population densities and likely to live in close proximity to livestock, poultry, companion animals and wild animals, in future, the state is likely to witness the emergence or re-emergence of new and existing zoonoses. This article examines public expenditure on zoonoses surveillance and control of vector-borne disease, size of livestock population, size of at-risk population who were employed in occupations that involved contact with animals and vectors and the cases due to zoonotic and vector-borne diseases in Kerala.

Public Expenditure on Integrated Disease Surveillance Programme and National Vector Borne Disease Control Programme

Recognising the public health implications of zoonotic diseases, the Government of India has allocated Rs 100 crore to surveillance, control and prevention of zoonotic and neglected tropical diseases in the budget 2018-19, which marks a doubling of allocation from the previous year. In addition to this, in 2018-19, the central government allocated Rs 4779 crore towards communicable diseases which includes National Vector Borne Disease Control Programme (NVBDCP). Table 17.1 presents information on actual expenditure by Kerala and the other South Indian states from the approved and available budget on Integrated Disease Surveillance Programme (IDSP) and National Vector Borne Disease Control Programme (NVBDCP) in 2016-17.

Table 17.1 : Share of approved and available budget spent by states on Integrated Disease Surveillance Programme (IDSP) and National Vector Borne Disease Control Programme (NVBDCP) in Kerala, Tamil Nadu, Karnataka, Andhra Pradesh, Telangana and All-India, 2016-17 (as %)

States	Integrated Disease Surveillance Programme		National Vector Borne Disease Control Programme	
	Expenditure from approved budget	Expenditure from available budget	Expenditure from approved budget	Expenditure from available budget
Kerala	88.1	72.8	73.8	56.4
Tamil Nadu	78.9	66.5	213.9	88.9
Andhra Pradesh	106.2	106.2	66.8	42.8
Telangana	90.6	75.2	71.7	47.2
Karnataka	88.4	74.4	96.7	72.6
All-India	80.5	65.1	80.9	50.6

Source: Financial Management Report 2016-17, National Health Mission

In the case of IDSP, among all the South Indian states, the share spent from approved and available budget was higher than the national average. However, expenditure from approved and available budget on IDSP was lower in Kerala than Andhra Pradesh, which spent more than the approved and available budget. Surveillance performs an important task in early identification, which then enables prompt public health responses. Clearly, the under utilisation of the budget for IDSP is a matter of concern in Kerala, which has witnessed outbreaks of zoonotic and vector-borne diseases like leptospirosis, scrub typhus and dengue. Further, expenditure from the approved and available budget on NVBDCP was lower in Kerala and Andhra Pradesh than the other South Indian states. This presents a challenge to strengthen efforts to control vectors at source.

In the 2018-19 budget, Kerala allocated Rs 352.4 crore to veterinary services and animal health, Rs 15 crore to insurance of livestock and poultry, Rs 4.8 crore to disease investigation among animals. While we do not have information from survey reports like National Sample Survey on the extent of insurance coverage of livestock and poultry population, the livestock census provides data on the size of the population.

The Size of Livestock, Poultry and Companion Animal Population in Kerala

The increasing number of some zoonotic and vector-borne diseases among human population highlights the need to focus attention on the health of livestock and companion animals, as incidence of some of these diseases among animals can have adverse effects on dairy, egg and meat production as well as cause disease among population engaged in animal-related occupations in the state. Kerala has witnessed a decline in livestock population between 2007 and 2012 (Ministry of Agriculture 2014) related to the shift in cropping pattern from paddy cultivation, which needed work animals to cash crops (Nair 1990). The move away from paddy resulted in a lowered demand for work animals (Nair 1990). In this context, we need to understand the size of livestock, poultry and companion animal population in Kerala.

Information from the 19th Livestock Census 2012 showed that there were 27.35 lakh livestock, 242.81 lakh poultry, 9.23 lakh dogs, 4681 stray cattle, 2.68 lakh stray dogs and 2.30 lakh rabbits in Kerala (Table 17.2).

Table 17.2 : Number of livestock, poultry, dogs and veterinary institutions in Kerala, district-wise, 2012

District	Livestock	Poultry	Dogs	Rabbits	Stray cattle	Stray dogs	Veterinary institutions
Kasaragod	120687	685297	40119	3509	1189	9331	42
Kannur	161706	798259	47897	8526	135	14753	91
Wayanad	116603	404289	25658	13805	70	3520	28
Kozhikode	155899	1122398	32287	10565	166	15956	89
Malappuram	239112	3276239	20151	24429	455	15198	110
Palakkad	293467	3512994	69553	6732	157	26772	100
Thrissur	256826	3185833	81344	19844	210	32833	118
Ernakulam©	248722	4270568	100405	22558	468	16647	105
Idukki	206222	527015	61047	39697	1566	14753	61
Kottayam	189875	1509805	77864	24106	54	10924	85
Alappuzha	137652	1532185	71063	11954	50	24354	81
Pathanamthitta	119119	643571	65748	8802	0	11548	60

(Contd...)

Kollam	220402	1070243	87055	12688	49	49582	83
Thiruvanantha-puram	268870	1743232	143168	23329	112	31749	99
Kerala	2735162	24281928	923359	230550	4681	268944	1152

Note: Total livestock covers cattle, buffalo, sheep, goat, pig, horses& ponies, mules, donkeys, camels, mithun and yak and total poultry include total birds in the poultry farms and hatcheries; Data on veterinary institutions is for 2011; Veterinary institutions include District Veterinary Centres, veterinary hospitals, veterinary dispensaries, veterinary polyclinics, mobile veterinary hospitals and mobile veterinary dispensaries

Source: Ministry of Agriculture, 2014, 19th Livestock Census, 2012

It is clear that poultry constitutes a large share of animals reared by people. Further, around 43 percent of poultry are raised in backyard poultry production, which means that these birds live in close proximity to human population (Ministry of Agriculture 2014). Nearly 19.5 lakh households in Kerala keep backyard poultry production (Ministry of Agriculture 2014). Among the livestock population in Kerala, cattle (48.6%) and goats (45.6%) accounted for a larger share.

Livestock population was highest in Palakkad, Thiruvananthapuram and Thrissur; poultry population in Ernakulam, Palakkad and Malappuram; dogs in Thiruvananthapuram, Kollam and Kottayam; stray cattle in Idukki and Kasaragod, and stray dogs in Kollam, Thrissur and Thiruvananthapuram. Malappuram and Thiruvananthapuram had the largest number of households with backyard poultry (Ministry of Agriculture 2014). Districts that had the largest number of veterinary institutions in the state were Thrissur, Malappuram, Ernakulam and Palakkad, which more or less corresponds to the size of livestock population and poultry.

Workforce in the Animal Husbandry, Meat Processing, Leather Tanning and Veterinary Activities

Information on sectoral contribution to Net State Domestic Product (NSDP) illustrates the meagre and declining share of the agricultural sector in Kerala (Table 17.3). In 2011-12, less than 10 percent of Kerala's NSDP was from agriculture and allied sector, which is much lower than the All-India share. The highest contribution to the state's NSDP was from the services sector (72%). Among the South Indian states, only Tamil Nadu (8.7%) has a similar pattern of low share of agriculture and allied sector to NSDP (Government of Tamil Nadu 2017).

Table 17.3 : Sector-wise share of Net Domestic Product in Kerala and All-India, 2004-05 to 2011-12 (as %)

Sector	Kerala		All-India	
	2004 - 05	2011 - 12	2004 - 05	2011 - 12
Agriculture and allied	16.5	8.5	19.9	14.7
Industry	22.3	19.5	25.2	24.1
Services	61.2	72.0	54.9	61.2
Total	**100**	**100**	**100**	**100**

Source: Government of Tamil Nadu, 2017

The data on share of agricultural households among rural households will show the extent to which agriculture remains important in Kerala. Information from National Sample Survey Office (2015) reveals that Kerala (27.3%), Tamil Nadu (34.7%) and Andhra Pradesh (41.5%) had the least share of agricultural households among rural households and Rajasthan (78.4%), Uttar Pradesh (74.8%) and Madhya Pradesh (70.8%) had the highest share. Kerala (34.5 %), West Bengal (30.3%) and Tamil Nadu (43.2%) had a relatively lower share of agricultural activities (cultivation and livestock activity) in the total average monthly income per agricultural household (NSSO 2015). In Kerala (83.9%), West Bengal (91.5%), Jharkhand (86%) and Bihar (85.3%) had a higher share of agricultural households which possessed land holdings of less than one hectare (NSSO 2015). Clearly, there has been a shift away from the agricultural sector in Kerala, which makes it more important to preserve the workforce who depend on it. Figure 17.1 shows that agricultural

Figure 17.1: Activities of agricultural households in Kerala and All-India, 2012-13 (as %)

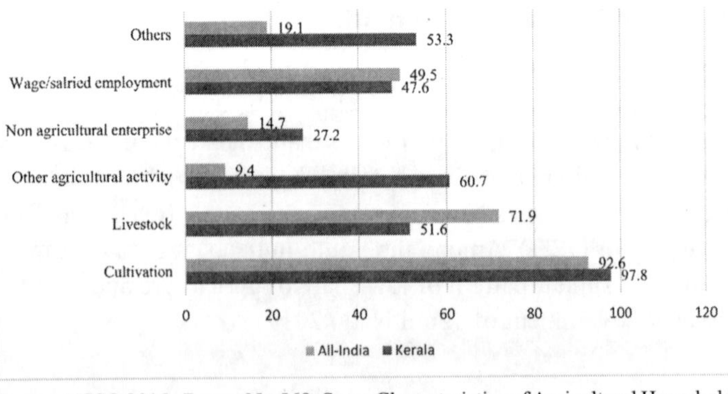

Source: NSSO 2015, Report No.569: Some Characteristics of Agricultural Households in India, January-December 2013

households in Kerala undertake cultivation (92.6%) along with other economic activities like working for salary elsewhere (47.6%), livestock rearing (51.6%), other agricultural activity (60.7%), non agricultural enterprise (27.2%) and other activities (53.3%), attesting to the importance of non farm employment.

The fact that nearly half of agricultural households were engaged in livestock rearing, points to the need for ensuring that animal health care practices are adopted so that animals and people who work with them remain healthy. The economic shocks to agricultural households that result from animal disease and human ill health, can lead to lower productivity and fatalities. With Kerala being one of the leptospirosis endemic states, and as it is transmitted from cattle, rodents and dogs to humans, efforts must be taken to prevent this disease. While cases of anthrax and brucellosis have not been documented recently in the state, surveillance efforts must be strengthened, as anthrax among cattle and humans have been reported from Andhra Pradesh. In 2018, 13 people were hospitalised for suspected anthrax in Andhra Pradesh (https://www.ndtv.com/andhra-pradesh-news/13-with-suspected-anthrax-infection-hospitalised-near-visakhapatnam-in-andhra-pradesh-1843513).

It is not just people engaged in livestock rearing who are at risk of contracting zoonotic and vector-borne diseases, but also those who work as farmers, agricultural labourers, in the animal slaughter and meat processing sector, in forestry and logging, hunting of animals, wholesale business of live animals, leather tanning industry and veterinary activities. Cultivators and agricultural labourers are exposed to infections from farm animals, ticks and from mosquitoes,which breed in stagnant water collected for irrigation. Workers who are in direct contact with live or dead animals are vulnerable to contracting diseases from these animals and insects which thrive on these animals like ticks and mites. For instance, a person who works in the animal husbandry sector and is involved in the maintenance of these animals can get diseases like anthrax, brucellosis, leptospirosis, avian influenza as well as tick- and mite-borne infections like scrub typhus and Crimean Congo Haemorrhagic Fever (CCHF). People employed in the slaughter of animals, meat processing industry and leather tanning industry are exposed to the same risk as those in animal production. Those engaged in forestry and logging face the risk of animal bites and tick-borne diseases like scrub typhus, Lyme disease and Kyasanur Forest Disease (KFD) as well as mosquito-borne diseases like malaria and dengue which thrive in dense vegetation. People employed in the veterinary sector are in direct contact with animals in the course of treating them and are vulnerable to acquiring diseases from infected animals if they do not observe safety protocols.

The Census of India 2011 shows that in Kerala, 1.18 lakh people (75,591 main workers and 42,681 marginal workers) were directly working in animal production like raising of cattle, buffaloes, sheep and goats, pigs, poultry and other animals and another 40,078 people in forestry and logging (Table 17.4).

Table 17.4 : Industrial classification of main and marginal workers in work related to animal husbandry, forestry, wholesale market for live animals and veterinary activities in Kerala, across social group, 2011

Industrial category	General			Scheduled Caste			Scheduled Tribe		
	Main worker	Marginal worker	Total	Main worker	Marginal worker	Total	Main worker	Marginal worker	Total
Cultivators	544932	125321	670253	16386	6422	22808	16288	4605	20893
Agricultural labourer	919136	403714	1322850	212417	86849	299266	65509	30520	96029
Animal production	75591	42681	118272	4185	2549	6734	1299	918	2217
Support activities for animal production	102	42	195	33	25	58	2	0	2
Hunting trapping & related activities	222	42	264	30	6	36	4	0	4
Forestry & logging	31035	9043	40078	6033	1880	7913	2601	1602	4203
Processing and preserving of meat	4349	1473	5822	96	58	154	18	14	32
Wholesale of agricultural raw materials and live animals	14003	1931	15934	205	72	277	52	16	68
Tanning and dressing of leather	4674	691	5365	367	105	472	8	0	8
Veterinary activities	4447	217	4664	511	19	530	67	0	67

Source: Census of India 2011

There were 13.2 lakh agricultural labourers and 6.7 lakh cultivators in the state. In Kerala, 15,934 people worked in wholesale of raw materials and live animals. Relatively smaller number of people were engaged in hunting of animals and leather tanning. Less than 5000 people were employed in veterinary services. Among Scheduled Castes (SCs), there were 2.9 lakh agricultural labourers, 22,808 cultivators, 7913 employed in forestry and logging and 6734 workers in animal production. Scheduled Tribes (STs) were mainly found in forestry and logging (4203) and animal production (2217). There were 96,029 agricultural labourers and 20,893 cultivators among STs. Among all workers in the state, 21.8 lakh were

Table 17.5 : Industrial classification of main and marginal workers in work related to animal husbandry, forestry, wholesale market for live animals and veterinary activities in Kerala, district wise, 2011

District	Industrial classification									
	Cultivators	Agricultural labourer	Animal production	Support activities for animal production	Hunting trapping & related activities	Forestry and logging	Processing and preserving of meat	Wholesale of agricultural raw materials and live animals	Tanning & dressing of leather	Veterinary activities
Kasaragod	23732	38774	3032	22	-	1838	116	326	164	132
Kannur	36549	82988	9087	16	16	3269	291	1339	731	261
Wayanad	52759	101630	6796	8	6	947	138	431	58	137
Kozhikode	26543	64198	8821	6	24	2503	480	2545	469	346
Malappuram	57567	117405	6702	14	24	4294	719	2150	259	336
Palakkad	67805	249949	8776	20	89	4848	480	1599	1305	329
Thrissur	41786	73941	12747	9	36	4068	741	1696	801	562
Ernakulam	50423	71391	9733	6	33	3867	731	1573	762	473
Idukki	100917	145698	5208	4	-	1608	136	480	30	186
Kottayam	49991	63052	15620	17	3	3310	489	1317	159	271
Alappuzha	25237	71706	6376	38	9	2206	296	464	421	275
Pathanamthitta	41414	55455	5230	10	10	1624	88	198	26	164
Kollam	56693	92193	7912	12	9	2782	348	1013	66	529
Thiruvananthapuram	38837	94470	12232	13	5	2914	769	803	114	663
Kerala	670253	1322850	118272	195	264	40078	5822	15934	5365	4664

Note: Workers include main and marginal workers; Processing and preserving of meat includes mutton-slaughtering and preparation, beef-slaughtering and preparation, pork-slaughtering and preparation, poultry and other slaughtering, preservation, processing and canning of meat, production of hides and skins originating from slaughterhouses, rendering of lard and other edible fats of animal origin, production and processing of animal offal, production, processing and preserving of other meat and meat products

Source: Census of India 2011

engaged in occupations that exposed them to risk of contracting zoonotic and vector-borne diseases. In the event of these at-risk workers contracting these diseases, the situation of informal workers are precarious. Informal workers like agricultural labourers, slaughter house workers, loggers etc. do not have access to paid leave in case they fall ill, which leads to loss of wages for the duration of their treatment. In addition, if they do not have access to subsidised public healthcare, expenditure on treatment can adversely affect their household income. Moreover, if the disease can transmit from human-to-human, then their household members and neighbourhood members are likely to contact it, since population density is very high in Kerala.

Which are the districts that have a larger number of workers who are vulnerable to zoonotic and vector-borne diseases? Table 17.5 reveals that Idukki, Palakkad and Malappuram had the highest number of cultivators. The number of agricultural labourers were highest in Palakkad, Idukki, Malappuram and Wayanad. Kottayam, Thrissur and Thiruvananthapuram had the highest number of workers employed in animal production. The largest number of workers in forestry and logging were in Palakkad, Malappuram and Thrissur.Workers involved in animal slaughter and meat processing were concentrated in Thrissur, Ernakulam and Malappuram. A larger number of people working in wholesale of raw materials and live animals were found in Kozhikode, Malappuram and Thrissur. Thiruvananthapuram, Thrissur and Kollam had a larger share of people employed in veterinary services.

Animal and Vector-Borne Diseases in Kerala

Data from the IDSP Annual Report 2017 reveals that routine surveillance had detected highest number of cases of leptospirosis in Kerala, Karnataka, Tamil Nadu and Maharashtra (NCDC 2017). The National Health Profile presents information on some vector-borne and zoonotic diseases like malaria, dengue, chikungunya, Japanese Encephalitis (JE) and rabies. Kerala along with Tamil Nadu has a large number of dengue cases (Table 17.6). This draws attention to the need for focussed state action on vector control and to utilise central allocation efficiently. In the case of other vector-borne diseases like malaria and chikungunya, the number of cases is lower in Kerala when compared to other South Indian states. But there have been three rabies deaths in Kerala. While rabies can be transmitted through dogs, cats, bats, wild animals etc., in India most cases are a result of contact with infected dogs.

In 2015, there were 1.25 lakh dog bite cases in Kerala (Deccan Chronicle 2016). A study in Kozhikode revealed that among all the animal bite cases, more than three-fourth of cases were bitten by dogs, 11.9 percent by fox, 7.6 percent by mongoose and 3.4 percent by wild pigs (Ramakrishnan *et al* 2017). Among those who reported dog bites, 97 percent were bitten by stray dogs (Ramakrishnan *et al* 2017). This suggests that there has been inadequate government attention to preventive health like animal birth control and garbage disposal in public spaces, which encourages stray dogs as well as vaccination of companion animals (like dogs and cats).

Table 17.6 : Cases and death due to select vector-borne and zoonotic diseases in Kerala, Tamil Nadu, Karnataka, Andhra Pradesh, Telangana and All-India 2017 (as absolute numbers)

State	Malaria		Dengue		Chikun-gunya	Japanese encephalitis		Rabies
	Cases	Death	Cases	Death	Cases	Cases	Death	Deaths
Kerala	1194	2	19973	37	78	1	0	3
Tamil Nadu	5449	0	23294	65	120	127	2	3
Andhra Pradesh	16913	0	4844	0	1147	1	0	0
Telangana	2688	0	3083	0	663	11	0	0
Karnataka	6529	0	17265	5	32170	23	0	15
All India	842095	104	157996	253	63679	2180	252	97

Note: Death due to Chikungunya are not reported in India
Source: CBHI 2018, National Health Profile

Nearly two-fifth of households in Kerala reported experiencing severe problem of mosquitoes and flies when compared to 62.4 percent in Andhra Pradesh, 47.4 percent in Karnataka and 59.1 percent in Tamil Nadu (NSSO 2014). Vector control measures by the state was inadequate in Kerala when compared to other South Indian states. Only 16.5 percent of households in Kerala reported that government agencies took efforts to control problem of mosquitoes and flies when compared to 54.5 percent in Tamil Nadu, 34.6 percent in Andhra Pradesh (undivided) and 29.1 percent in Karnataka (NSSO 2014).

Temporal and Regional Variations of Zoonotic and Vector-borne Diseases in Kerala

The IDSP data presents information on zoonotic and vector-borne diseases like leptospirosis, scrub typhus, Kyasanur Forest Disease, Japanese Encephalitis (JE) malaria, dengue and chikungunya. The number of leptospirosis cases and deaths have increased by 125 percent and 158 percent respectively from 2013-2018 in Kerala (Table 17.7). After the natural disaster in Kerala in 2018, there was an outbreak of leptospirosis, which is a bacterial infection transmitted by rodents and other animals such as cattle and dogs to humans. The close links that leptospirosis has with floods is established when we examine the month-wise data on leptospirosis cases in Kerala. The number of leptospirosis cases increased from 80 in May to 247 in August and reached a peak of 856 in September 2018, coinciding with the floods (http://dhs.kerala.gov.in/pdf2018/data_24112018.pdf). The increase in leptospirosis cases also point to another concern - livestock and companion animals' health, as it affects these animals as well. Scrub typhus is spread to humans through the bites of infected chiggers. While the cases due to scrub typhus have increased between 2013 and 2018, it can be seen that there was a decline after 2015.

Table 17.7: Cases and death due to select vector-borne and zoonotic diseases in Kerala, 2013-18

Communicable diseases		2013	2014	2015	2016	2017	2018
Leptospirosis	Cases	814	1075	1098	1710	1408	1833
	Death	34	43	43	35	80	88
Scrub typhus	Cases	68	433	1149	633	340	226
	Death	0	6	15	3	5	4
Kyasanur Forest Disease	Cases	1	6	102	9	0	0
	Death	0	0	11	0	0	0
Japanese Encephalitis	Cases	2	3	0	1	1	5
	Death	0	2	0	0	0	2
Malaria	Cases	1634	1751	1549	1540	1177	789
	Death	0	6	4	3	2	0
Dengue	Cases	7938	2548	4114	7218	21993	3834
	Death	29	13	29	21	165	32
Chikungunya	Cases	247	264	152	124	54	52

Note: Data for 2018 is for January- October 2018

Source: http://dhs.kerala.gov.in/pdf2018/data_24112018.pdf

Information presented in Table 17.8 shows that there were inter-district variations in cases and deaths caused by zoonotic and vector-borne diseases across the districts in Kerala. The data points to higher cases of leptospirosis in Pathanamthitta, Kozhikode, Thiruvananthapuram and Alappuzha, indicating that it was dispersed across regions and not concentrated in a particular region. Deaths due to leptospirosis was higher in Kollam and Kozhikode. Pathanamthitta and Alappuzha were badly affected by the floods, which enabled the spread of leptospirosis. In the previous year also, the highest number of leptospirosis cases were reported from Thiruvananthapuram, Alappuzha and Kozhikode(http://dhs.kerala.gov.in/pdf2018/data2017dhs.pdf). Thiruvananthapuram has a high livestock and dog population, which can transmit leptospirosis.

Table 17.8: District-wise data on select zoonotic and vector-borne diseases in Kerala January- October 2018

District	Leptos-pirosis		Scrub typhus		Japanese Encephalitis		Malaria		Dengue		Chikun gunya
	Cases	Deaths	Cases	Deaths	Cases	Deaths	Cases	Deaths	Cases	Deaths	Cases
Kasaragod	28	2	3	-	-	-	136	-	634	3	-
Kannur	35	1	8	-	-	-	109	-	328	3	1
Wayanad	108	3	16	1	-	-	15	-	40	-	-
Kozhikode	243	12	25	-	4	2	87	-	237	3	-
Malappuram	183	5	5	1	-	-	84	-	822	9	-
Palakkad	93	7	6	-	-	-	22	-	281	7	-
Thrissur	80	11	1	-	-	-	68	-	194	-	32
Ernakulam	77	2	-	-	-	-	80	-	167	-	-
Idukki	30	3	3	-	-	-	21	-	51	-	-
Kottayam	106	1	7	-	-	-	24	-	114	1	1
Alappuzha	225	5	-	-	1	-	22	-	126	1	1
Pathanamthitta	280	9	4	-	-	-	32	-	356	2	-
Kollam	115	16	6	1	-	-	31	-	235	1	-
Thiruvananthapuram	230	11	142	1	-	-	58	-	249	2	17
Kerala	1833	88	226	4	5	2	789	0	3834	32	52

Source: http://dhs.kerala.gov.in/pdf2018/data_24112018.pdf

While the earlier outbreak of scrub typhus was concentrated in Wayanad, by 2018, the highest number of cases was found in Thiruvananthapuram (142) and Kozhikode (25). In 2017 the largest number of scrub typhus cases were reported from Thiruvananthapuram and Wayanad (http://dhs.kerala.gov.in/pdf2018/data2017dhs.pdf). Scrub typhus is usually associated with those living near, working in or visiting forested areas where chigger mites are found. So the large number in Thiruvananthapuram raises questions on whether there has been increase in economic activities in forests or whether chiggers have become resistant to insecticides used in agricultural farms. Further, a decline in chiggers' usual host population and availability of new hosts like humans could be a factor in the large number of cases of scrub typhus in Thiruvananthapuram. The few cases of JE were concentrated in Kozhikode district. While the number of cases due to malaria was higher in Kasaragod and Kannur, dengue cases were higher in Malappuram and Kasaragod. Thrissur had the largest number of chikungunya cases. In 2017, Thiruvananthapuram, Kollam and Palakkad had the highest dengue cases (http://dhs.kerala.gov.in/pdf2018/data2017dhs.pdf).

In addition to the size of the livestock population, regional variation in zoonotic and vector borne cases could be associated with population density, urbanisation, land use changes, changes in agricultural practices, rainfall, garbage disposal practices, urbanisation etc.

As Kerala has the third highest human population density (860 persons per sq km) after Bihar and West Bengal, people are likely to live in close proximity to livestock and poultry, companion animals and wild animals. Among all districts the highest human population densities were found in Thiruvananthapuram, Alappuzha, Kozhikode, Malappuram, Ernakulam, Kollam and Thrissur (Figure 17.2). Idukki, Wayanad and Pathanamthitta have the lowest human population densities.

Figure 17.2: Density of human population (per square km) in Kerala, district wise, 2011

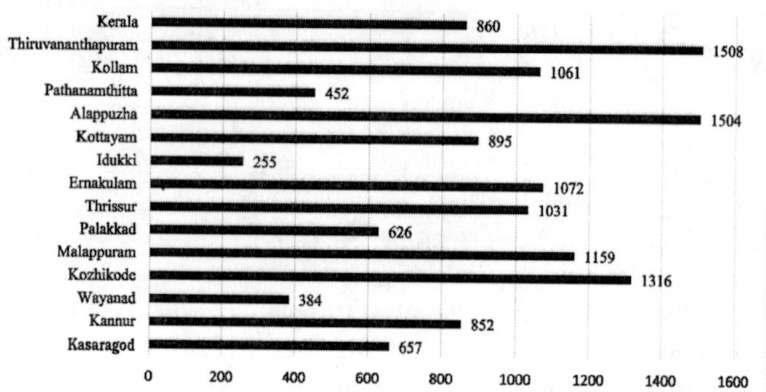

Source: Census of India, 2011, District Census Handbook, Kerala (http://www.censusindia.gov.in/2011census/dchb/DCHB.html)

High population densities combined with large livestock population increases the contact between humans and animals, which in turn increases the likelihood of zoonotic disease transmission if good practices in animal health and in handling of animals are not followed. Districts that have a high human population density, like Thiruvananthapuram, Alappuzha and Kozhikode also had a high number of leptospirosis cases. Thiruvananthapuram also had a high number of scrub typhus and dengue cases.

Data from the Census of India, 2011 reveals that Kerala was one of the most highly urbanised states in India after Tamil Nadu, with 47.7 percent of the population residing in urban areas. The districts with the highest proportion of population living in urban areas were Ernakulam (68.1%), Thrissur (67.2%) and Kozhikode (67.2%). Among these districts, only Kozhikode has a high number of people affected by zoonotic and vector-borne diseases. Notably, urbanisation increased by nearly 28 percentage points in Kozhikode between 2001 and 2011, which suggests that land use changes and high human density could be possible explanations.

The Way Forward in Rebuilding Kerala: Integrating Animal and Human Health with the Ecosystem

The floods of 2018 presents an opportunity to Kerala to develop an integrated approach to animal, human health and the ecosystem, which is called the One Health approach. The marked increase in zoonoses like leptospirosis and scrub typhus as well as vector-borne diseases like dengue during the floods and the period following it are not an unexpected phenomenon in the state. Leptospirosis, scrub typhus and dengue cases have witnessed an increase over a period of time pointing to anthropogenic and possibly natural changes to the ecosystem, which has brought animals, vector, pathogens and human hosts into close contact with each other. In rebuilding Kerala, it is imperative that the state takes efforts to control these zoonotic pathogens at their animal source and vector-borne diseases with a coordinated multisectoral approach across departments of health, agriculture, animal husbandry, environment, forest & wildlife, fisheries, water supply and sanitation and food safety. For instance, in the control of leptospirosis which is spread through rodents, dogs, cattle and wild animals, there has to be coordinated action among the ministries of health, animal husbandry, forest and agriculture.Vaccination is available for pet dogs and cats as well as livestock against leptospirosis. Some agricultural practices could be vector-friendly and hence before any large scale scheme is implemented, Centre for One Health Education Advocacy Research and Training (COHEART) and entomologists should be consulted for vector control. In the case of rabies and leptospirosis, the ministries of health, animal husbandry and Kerala Total Sanitation and Health Mission and local self government institutions will need to develop a strategy to register all pet animals like dogs and cats, ensure that pet animals are vaccinated, implement birth control and vaccination of stray dogs, and enforce waste management protocols across the state. As people who work in forestry and logging, government employment programmes and live near forests are vulnerable to scrub typhus, KFD and Lyme disease, the ministries of forest and wildlife, animal husbandry and Mahatma Gandhi National Rural Employment Guarantee Scheme under National Livelihood Mission will need to work out strategies to minimise tick and chigger bites among such workers. As backyard poultry comprises a large share of poultry production in Kerala, the department of animal husbandry and health have to educate people about correct and safe poultry rearing practices.

The Government of India has not given zoonoses and vector-borne diseases the attention that it deserved in health policy. The National Health Policy 2017 mentions only management of zoonotic diseases like rabies, need for surveillance of outbreaks and treatment protocols for vector-borne diseases. One Health approach to animal and human health does not find mention in the National Health Policy 2017. It has been noted that intersectoral mechanisms to operationalise One Health approach were not coordinated at the national level(Chatterjee *et al* 2016).

Kerala has taken a step in the right direction with establishment of Centre for One Health Education Advocacy Research and Training in 2014 at Wayanad under Kerala Veterinary and Animal Sciences University. However, this approach needs to be operationalised across sectors. The example of Bangladesh's 'Strategic framework for One Health approach to infectious diseases in Bangladesh' in 2012 offers some pointers.For the One Health approach to succeed in Kerala, it is important to involve communities in surveillance, changing unhealthy agricultural and animal husbandry practices, and protection of forest wealth with least economic disruptions to communities. Involving people in the process of integrating animal and human health is the way ahead.

References

1. Anish, TS, Vijayakumar, K and Itty Amma KR Leela (2011): "Domestic and Environmental Factors of Chikungunya-affected Families in Thiruvananthapuram (Rural) District of Kerala, India", *J Glob Infect Dis*. 3(1): 32–36.
2. Central Bureau of Health Intelligence (2018): "National Health Profile 2018", New Delhi: Ministry of Health and Family Welfare.
3. Chakravarti A, Arora R, Luxemburger C. (2012): "Fifty years of dengue in India", *Trans R Soc Trop Med Hyg*, 106: 273–282.
4. Chatterjee, P, Kakkar, M, Chaturvedi, S. (2016): "Integrating one health in national health policies of developing countries: India's lost opportunities", *Infectious Diseases of Poverty*, 5(87): 1-5.

5. Deccan Chronicle (2016): Kerala: Dogs bit 3.98 lakh people in last 3 years, Rohit Raj, November 5 2016.

6. Dubal Z B, Barbuddhe S B and Singh N P. (2014): "Important Zoonotic Diseases: Prevention and Control", *Technical Bulletin No. 39*, Goa:ICAR Research Complex for Goa.

7. George, P.S. and Chattopadhyay, S. (2001): "Population and Land Use in Kerala", In: *Growing Populations, Changing Landscapes: Studies from India*, China and the United States,Washington DC: National Academy of Sciences, 79-106.

8. Government of Tamil Nadu (2017): *Tamil Nadu Human Development Report 2017,* Chennai: Tamil Nadu State Planning Commission

9. https://www.ndtv.com/andhra-pradesh-news/13-with-suspected-anthrax-infection-hospitalised-near-visakhapatnam-in-andhra-pradesh-1843513

10. India Today (2017): Migrant workers face challenges in Kerala: Workshop (https://www.indiatoday.in/pti-feed/story/migrant-workers-face-challenges-in-kerala-workshop-939635-2017-06-07)

11. Ministry of Agriculture (2014): *19th Livestock Census, 2012,* Vol. 2, New Delhi: Department of Animal Husbandry, Dairying and Fisheries

12. Morse, S.S. (1995): "Factors in the emergence of infectious diseases", *Emerg. Infect. Dis*. 1, 7"15

13. Mutheneni, SR, Morse, AP, Caminade, C, Upadhyayula, SM. (2017): "Dengue burden in India: recent trends and importance of climatic parameters", *Emerging Microbes & Infections*, 6 : 1-10

14. Nair, KN. (1990): "Cattle Development in Kerala: Trends and Prospects", *Economic and Political Weekly*, 25 (35-36): 2005-2010.

15. National Centre for Disease Control (2017): *IDSP Annual Report 2017,* New Delhi: NCDC

16. National Sample Survey Office (2015): "Some Characteristics of Agricultural Households in India, January-December 2013", 70th Round, *Report No.569* (70/33/1), New Delhi: NSSO

17. National Sample Survey Office (2014): "Drinking Water, Sanitation, Hygiene and Housing Condition in India 2012" , 69th Round, July-December 2012, *Report No 556* (69/1.2/1), New Delhi: NSSO

18. National Sample Survey Office (2016): "Swachhta Status Report 2016", New Delhi: NSSO

19. Patz, JA, Graczyk, TK, Geller, N, Vittor, AY.(2000): "Effects of environmental change on emerging parasitic diseases", *International Journal for Parasitology*, 1-11

20. Ramakrishnan LV, Shakir L, Chandran P, Thavody J. (2017): "Profile of Mass Casualties of Animal Bite Cases Reported to a Tertiary Care Institution from April 2014 to March 2015 in North Kerala", *India. Int J Med. Public Health,* 7(1): 33-7

21. Satish, KV, Saranya KRL, Reddy CS, Hari Krishna P. Jha, CS, Prasada Rao PVV. (2014):"Geospatial assessment and monitoring of historical forest cover changes (1920–2012) in Nilgiri Biosphere Reserve, Western Ghats, India", *Environmental Monitoring and Assessment*

22. Sukumaran, A and Pradeepkumar A.S. (2015): "One Health approach: A platform for intervention in emerging public health challenges of Kerala state", *International Journal of One Health*; 1:14-25.
23. The New India Express (2018): "Man with Congo fever under treatment in Thrissur", December 4 2018 (http://www.newindianexpress.com/states/kerala/2018/dec/04/man-with-congo-fever-under-treatment-in-thrissur-1906815.html)
24. The Times of India (2013): "Lyme disease outbreak in Wayanad", Rajeev KR, March 2, 2013 (https://timesofindia.indiatimes.com/city/kozhikode/Lyme-disease-outbreak-in-Wayanad/articleshow/18758675.cms)

18

Employment and Livelihood in Kerala: Issues and Options

M. Suresh Babu

The incongruence between output growth and employment growth has been a problem with the Kerala's model of development. While the economy witnessed structural transformation since 1980-81, with a sharp reduction in the share of primary sector in GSDP the corresponding decline in employment share has not taken place in primary sector. Kerala did not experience a sequential growth process. Even though service sector emerged as the engine of growth it did not provide employment matching with its income. The question of employment generation got compounded as the process of industrialization failed to take off with the share of income from secondary sector not commensurating with the level of employment in the sector. Service sector's growth has led to a peculiar situation with the overall econnomic growth continuing to maintain a trend rate of around 6% with most of the commodity production sectors either declining or stagnating. This growth of service sector has also triggered the process of urbanization with the urban sector almost doubling itself during the last decade with the share of urban population shooting up from around 26% in 2001 to 48% in 2011, an 83% increase which is a historical high, and puts the share of urban population in the state roughly 50% larger than the national average.

Despite the structural transformation, the economy continues to be sluggish in generating enough jobs. The unabated decline in the agriculture sector and stagnant industrial sector leaves the only booming service sector as the prime employment creator. Unpacking the service sector shows that its growth rate is fuelled by a narrow base, that is, in terms of expansion of healthcare and transport facilities. While demographic factors such as increasing aged population and growing morbidity have contributed to the growth of health care sector, transport sector has grown due to higher public investments in roads. The lack of dynamism in the modern service sector casts a shadow on the sustainability of service sector growth and its ability to generate employment for educated youth in the state.

Structure of the Economy

The quick estimate of Gross State Domestic Product (GSDP) at constant (2011-12) prices during 2016-17 registered a growth rate of 7.41 per cent in 2016-17 compared to 6.60 per cent in 2015-16. In current prices the GSDP has registered a growth rate of 10.59 percent in 2016-17. Consequently the per capita GSDP at constant (2011-12) prices in 2016-17 was 140,107, registering a growth rate of 6.88 per cent in 2016-17. At current prices, the per capita GSDP in 2016-17 registered a growth rate of 10.04 per cent over the previous year. At constant (2011-12) prices, per capita state income has been higher than the per capita national income. Thus in terms of the growth of the economy we find that there exists a rate of growth that is higher than the national average. The arguments of a paradox of higher rate of developemnt co-existing with lower levels of growth does not seem to hold good for Kerala.

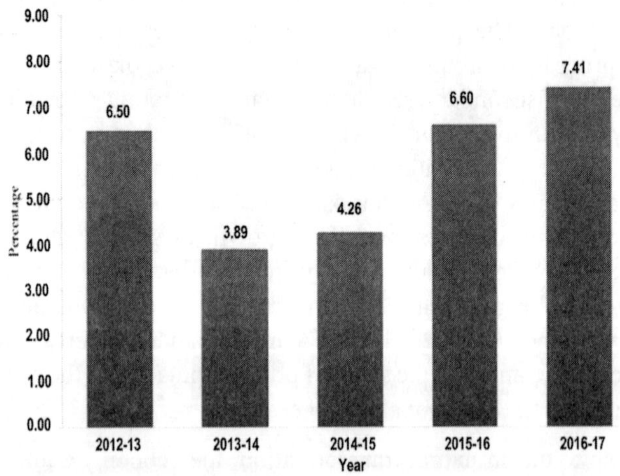

Source: Kerala state planning board, Economic Review 2017

Figure 18.1. GSDP Growth Rate

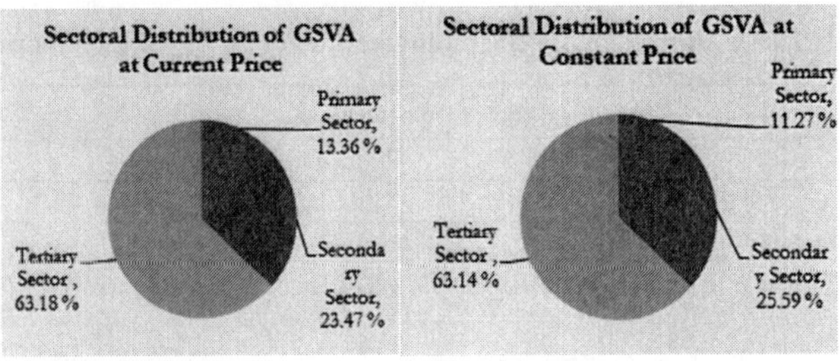

Source: Kerala state planning board, Economic Review 2017

Figure 18.2: Sectoral Distribution of Valued Added

The higher rate of growth of the economy is fuelled by an expanding service sector. Sectoral decomposition of gross state value added (GSVA) shows that the contribution to GSVA of the tertiary sector increased from 62.59 per cent in 2015-16 to 63.18 per cent in 2016-17. During the corresponding period the contribution of secondary sector declined from 24.59 per cent to 23.47 per cent. An analysis of annual sectoral growth rate of GSDP shows that tertiary sector recorded the highest rate of growth of 6.7 per cent in 2016-17. Interestingly during the same period, the growth rate in primary sector increased from (-) 11.2 per cent to 5.19 percent. This is mainly because of the increase in production of some of the crops, live stock, fishing and aquaculture. However, if we examine a slightly longer time period we find that during the period 2013-14 to 2015-16, the share of tertiary sector has increased from 60.43 to 63.66 per cent and the share of primary sector decreased from 13.45 per cent to 12.07 per cent and secondary sector from 25.81 to 24.27 per cent.

Employment

Given the specatcular growth performnace of the economy, it is expected that the rate of growth of employment would be at a rate that would reduce the unemployment in the economy. This assumes significance as the service sector which is the dominant sector also comprises of a'modern services sector' apart from the traditional sectors, which generates activities that provide employment for the educated work seekers. Given the preponderance of educated employment seekers in Kerala, growth of service sector is thus viewed as a favorable situation. The evidence however, is to the contrary.

Table 18.1 : State/UT wise unemployment rate based on UPS Approach

Andhra Pradesh	3.9	Telangana	2.8
Karnataka	1.5	Puducherry	4.9
Kerala	12.5	All India	5
Tamil Nadu	4.2		

Source: Fifth Annual Employment-Unemployment Survey (2015-16), Govt. of India.

In table 18.1 presents the unemployment rate in Kerala and other southern states in comparison with the all India levels. It is striking to note that Kerala has the highest unemployment rate, based on UPS approach in the southern region. In fact it has crossed double digits at 12.5 percent, compared to that of 1.5 percent in Karnataka which is the lowest rate in the region. All the southern states have unemploymnt rates less than the national rate of 5 per cent. Puducherry's rate which is the highest among states in the southern region excluding Kerala is less than half of Kerala's rate. This clearly shows that the generation of jobs is just not enough to absorb the increasing work force in Kerala.

Table 18.2 : Employment in the Organised Sector in Kerala (Lakh Persons as on 31st March)

	Men	Women	Total
2010	6.67	4.48	11.15
2011	6.66	4.45	11.11
2012	6.4	4.49	10.89
2013	6.31	4.57	10.88
2014	6.52	4.77	11.29
2015	5.74	5.62	11.36
2016	6.83	5.02	11.85
2017	6.64	5.09	11.73

Source: Kerala state planning board, Economic Review 2017

Viewed from a slightly different perspective, it is often argued that the generation of more jobs in the organised sector is an indicator of better quality jobs in an economy. Organised sector provides social security and better working conditions than the unorganised or informal sector. Thus workers in this sector enjoy better previleges and job security vis-à-vis workers in the informal sector. Growing sectors in an economy at times generates low quality jobs, which might reduce the unemployment, but would persist the vulnerablities. From table 18.2 it is evident that the rate of growth of employment in the organised sector in Kerala has been only marginal since 2010. In fact we find that in a number of years (four years) the absolute level of employment in organised sector declined. One positive trend that is visible is that the number of women workers in organised sector has increased marginally over the years. Thus it is evident that not only the unemployment rate is high in Kerala, but also the extent of jobs in the organised sector has also grown at a slower pace, indicating the low quality of jobs being generated.

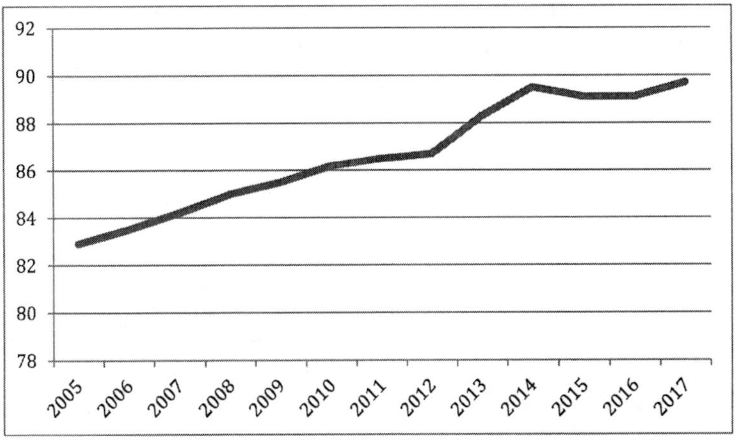

Source: Kerala state planning board, Economic Review 2017

Figure 18.3: Percentage of work seekers in Kerala by Educational Level SSLC & above

The supply of jobs have to be matched by demand for jobs in terms of the skills and preferences of the job seekers. In this context Kerala faces a unique situation as the jobseekers are more educated compared to other states in India. Figure 18.3 reveals that almost 90 percent of work seekers in Kerala have educational qualifications of SSLC and above. This percentage has grown over time. We find that in 2005 the percentage of work seekers with qualifications above SSLC was 82 which has increased by nearly 10 percentage points in the last 10 years. This poses two challenges for the economy. First, the economy has to generate sufficient quantity of jobs to absorb the increasing workforce and second, the kind of jobs generated needs to match the higher skills acquired by the work seekers. The structural change that has taken place in the economy has to accommodate these two issues. We find that in Kerala's case both the challenges remain to be addressed.

Shrinking Production Base

The question of employment and livelihood is related to use of resources in the state. The changes in land use pattern as protrayed in table 18.3 in Kerala were unprecedented during the past decades in terms of deforestation, increase in area as current fallow, increase in area under non-agricultural land, decrease in both net area sown and gross cropped area resulting in decline in cropping intensity. The declining cultivable area, predominance of tiny and fragmented holdings, decline in work force in terms of reduction in agricultural labour and cultivator has made farming more vulnerable.

Table 18.3: Land use pattern
Land use Pattern in Kerala (Area in Ha)

Sl.No	Classification of Land	2015-16	2016-17	Percentage of Geographical areas	Change in area between 2015-16 and 2016-17	
					Actual	Percentage
1	Total Geographical Area	3886287	3886287	100	0	0
2	Forest	1081509	1081509	28	0	0
3	Land put to non-agricultural uses	434646	441934	11	7288	2
4	Barren and uncultivated land	13100	11780	0.3	-1320	-10
5	Permanent Pastures and Grazing land	0	0	0	0	0
6	Land under miscellaneous tree crops	2663	2450	0.06	-213	-8
7	Cultivable waste	99499	101379	3	1880	2
8	Fallow other than current Fallow	55258	55530	1	272	0.49
9	Current Fallow	70003	72008	2	2005	3
10	Net area sown	2023073	2015482	52	-7591	-0.38
11	Area sown more than once	604504	568518	15	-35986	-6
12	Total cropped Area	2627577	2584000	66	-43577	-2
13	Cropping Intensity (%)	130	128	0		

Source : Directorate of Economics and Statistics

The stagnation in production sectors of the economy, which has contributed to low absoprtion of workers is evident from the case of agriculture, the most recent trend of which is shown in table 18.4. In agriculture, area under production of food grains has declined by more than half from 3.3 lakh hectares to 1.6 lakh hectares between 2000-01 and 2014-15 while that of coconut has fallen from 9.3 lakh hectares to 7.9 lakh hectares. Area under major spices - like pepper, cardamom and ginger and also tea - has also declined. Similarly, the number of livestock has fallen from 34.8 lakh to 27.3 lakh during the period.

Table 18.4 : Area, Production and Productivity of Principal Crops

Sl.No.	Crops	Area(Ha)		Production(T)		Productivity (Kg /Ha.)	
		2015-16	2016-17	2015-16	2016-17	2015-16	2016-17
1	Rice	196870	171398	549275	437112	2790	2550
2	Pulses including Tur	3764	1738	4265	1711	1133	984
3	Pepper	85948	85207	42132	34065	490	400
4	Ginger	4986	5147	22044	20478	4421	3979
5	Turmeric	2603	2631	7112	6506	2732	2473
6	*Cardamom	39730	39080	19500	17147	491	439
7	Areca nut	99126	97696	132453	116839	1336	1196
8	Banana	59835	57157	536155	489322	8961	8561
9	Other Plantains	57683	57138	411626	395806	7136	6927
10	Cashew nut	43090	41660	24733	27944	574	671
11	Tapioca	69405	68664	2662610	2529729	38363	36842
12	**Coconut	790223	781495	5873	5379	7432	6883
13	***Coffee	84987	84976	69230	63476	815	747
14	$Tea	30205	30205	57898	61505	1917	2036
15	# Rubber	550840	551050	438630	540400	796	981

***Production in Million Nuts, Productivity in Nuts/Ha*
** Spices Board, # Rubber Board, *** Coffee Board, $ Tea Board*
Source: Directorate of Economics and Statistics

One Possible Option

Given this context the recent natural calamity in Kerala raises further questions on the economy's ability to generate jobs and provide stable livelihood. However, it also provides an opportunity to recast the trajectory of development. A possible area to explore in this context would be forge stronger links between agriculture and industrial sectors. Agriculture alone will not be sufficient to address the questions of poverty and inequality. It is becoming increasingly crucial to focus immediate attention on agro-industries. Such industries, established along efficient value chains, can increase significantly the rate and scope of industrial growth. Agro-industrial products offer much better prospects of growth than primary commodities. In addition, the marked trend to break down production processes into specific tasks opens up new opportunities for Kerala to specialize and take a more profitable part in global trade provided we meet increasingly stringent market requirements.

A significant proportion of state funds are now used to support agricultural production inputs - primarily seeds, fertilizers and irrigation systems. Little attention has been paid to the value chains by which agricultural products reach final consumers and to the intrinsic potential of such chains to generate value added and employment opportunities. However, participation in value chains implies both opportunities and pitfalls for Kerala. The prospect that lead firms such as brand owners, innovators and system integrators may appropriate increasing shares of rent and therefore further widen the gap is very real. Another danger is that SMEs (including farmers) will face difficulties as international supermarkets, retailers and buyers govern the access to markets by setting up food safety and quality standards that impose a substantial cost of compliance.

One opportunity resides in the fact that, in pursuit of lower transport and transaction costs, TNCs increasingly subdivide and relocate production processes in developing regions. In addition, modern chains require smooth product flows, high standards and error-free production. Consequently, lead firms are willing to invest in knowledge transfer to the benefit of local industries, institutions and service providers. Other opportunities exist in niche markets, both domestic and foreign, particularly for specialized and natural products found only in Kerala. Small farmers can also participate in these chains that yield higher incomes. Given the present context in Kerala there exists a need to explore these new options, learning from the development experience of other emerging economies.

19

The Implications of Disaster on Dynamics of Migration in Kerala

K. Jafar

1. Introduction

The state of Kerala is known for its unique development experiences; the existing socio-political context and State's role in prioritising the social sector created a strong human capital base in the state. This enabled its educated workforce in exploring the new options of livelihoods which emerge outside the state and large-scale migration of workers. Kerala has long history of labour migration; its experience during the last four decades is particularly known for its dynamic nature and impact on the social, political and economic lives in the region. While considering the differential impact, Gulf-migration found to be more effective in addressing the vulnerabilities and in improving the social and economic mobility among the lower and middle level social and economic groups. Migration affects almost all aspects of Malayalee lives; it has helped the Kerala society in addressing several issues while it brought some fresh ones in. Advance in communication technologies enables the global Malayalee diasporas in following the every-day lives in Kerala closely; their participation and contribution remain as the key source of support in almost all collective efforts made in Kerala. Their support in terms of cash and kind has been very critical in managing the recent disaster especially in the relief and rehabilitation process in different parts of the state. Among others, the support offered by the rulers, officers and the larger public in Gulf-countries were remarkable and the same can be seen as a gesture of strong relationship with those countries and recognising the contribution of Malayalee workers in developing these countries.

The state recognises these factors and makes all effort in its appeal to the global Malayalee diasporas for supporting its efforts in building New Kerala in a sustainable manner.

The paper makes an attempt to look in to the nature of recent disaster and its implications on issues related to labour migration and livelihood in the post-disaster context in the state. While doing this, it also tries to reflect on some key aspects of employment and larger process of development in the state. Given the nature of information available on key indicators, the paper uses the available base-line data sets on nature of employment, labour migration and explores the available information on how the disaster affected these patterns.

2. The Nature of Disaster

Between June 1 and August 18, 2018, Kerala experienced the worst ever floods in its history since 1924. The estimates indicate that Kerala received an excess rainfall of 96 percent during August (1-30 August 2018), and 33% during the entire monsoon period till the end of August (Government of Kerala, 2018). The heavy rainfall received during June-August months triggered several landslides and flood in different parts of the state. Though the existing mechanisms and institutional arrangements were not fully prepared to manage the situation in the early phase, state managed to mobilise all its resources and organised the support from all possible corners in managing the crisis at a critical stage and controlled the casualty to the minimum. Even with all these efforts, estimates highlight the loss of lives and properties in different parts of the state. According to latest reports by the Government of Kerala, 1259 out of 1664 villages spread across its 14 districts were affected.[1] Some districts like Alappuzha, Ernakulam, Idukki, Kottayam, Pathanamthitta, Thrissur and Wayanad were severely affected. It affected 5.4 million people, displaced 1.4 million people, and took 483 lives (from May 22 to August 29). The post-disaster need assessment (PDNA) estimates total damage around Rs. 10,561 crore and losses around Rs. 16159 crore i.e. a total disaster effects of around Rs.26718 crore without including the damage estimates from the Joint Rapid Damage and Needs Assessment (JRDNA) conducted by the World Bank and the Asian Development Bank. When the JRDNA estimates are included, the total recovery needs can be estimated around Rs. 31000 crore (PDNA, 2018). Many had to live in the temporary shelters and relief camps for a longer duration. The disaster caused severe damage to houses, infrastructure especially electricity, roads and public institutions that provides basis services and washed away crops and livestock. Its impact on Kerala society and economy particularly on state finance and productive sectors will be very crucial. This will have direct impact on the lives and livelihoods of people in the affected areas and the same need not be limited to the existing estimates.

3. Impact of Mass-Education on Occupational Patterns

Kerala's development experience can be seen as a good example to show how the state's strategy of prioritising the social sector to strengthen the process of human development and the same helping the state in generating economic growth in the state. One can identify the 'Kerala model' and 'virtuous' growth as two distinct but related growth phases; the high growth phase experienced in the late 1980s can be identified as a result of specific strategy followed in the earlier period. It helped the state in strengthening the mass-education process and enabled its educated workforce in seeking opportunities outside the state. Later, migration and inflow of remittances fueled the domestic economy and affected the process of development in the state (Kannan, 2005 & 2007; Jafar, 2018; Zachariah, Mathew, & Rajan, *2003;* Zachariah, & Rajan, 2012 & 2013). The larger trend in employment in Kerala indicates a low rate of labour force participation (includes employed and unemployed persons as a share of total population) for many years (40%, 57.5% and 24.0% respectively for total, men and women in 2012). The workforce participation rate (employed as a share of total population) indicates that there has been some improvement in recent years while the improvement was minimal among the women. Within the state, districts like Palakkad, Malappuram, and Kasaragod etc. experienced marginal decline in women's work participation rate (Census of India, 2011). The relatively lower participation in labour market particularly among the women can be linked to the progress in education and emigration of workers (mainly men) from Kerala.

Table 19.1: Distribution of Main Workers in Kerala 1991-2011 ***(Share %)***

Occupational Category	**1991**	**2001**	**2011**
Cultivators	12.2	7.1	5.8
Agricultural Labourers	25.5	12.4	9.9
Livestock, Forestry, Fishing, Hunting, plantation, Mining and Quarrying etc	10.2	13.8	10.8
Household Industry	2.6	3.3	2.1

(*Contd...*)

Other than Household Industry*	11.6	12.2	12.4
Construction	4	9.3	14.3
Trade and Commerce**	12.6	14.9	14.2
Transport, Storage and Communication	6	9.2	9.0
Other Services***	15.2	17.2	21.6

Note: *includes Electricity, gas, and water supply; ** includes wholesale and retail trade, hotels & restaurants;

*** includes financial intermediaries, real estate activities and categories from 'L' to 'Q' according to National Industrial Categories 2001 (include HHI from other service sectors)

Source: Census of India

The dynamics of education and migration affected the larger pattern of transition; it has led to emergence of 'non-agrarian' villages in the state (Pani, & Jafar, 2010). It seems to have played a key role in changing the occupational pattern of workers in the state. We notice steady decline in share of main workers engaged in traditional jobs in primary sectors while this has been followed with gradual increase in the share of workers engaged in jobs available in the service sector. The share of main workers engaged as cultivators declined from 12.2 per cent of 1991 to 7.1 per cent in 2001 and further declined to 5.8 per cent in 2011, the share of agricultural labourers declined from 25.5 per cent to 12.4 per cent and 9.9 per cent during the same periods. On the other hand, share of construction workers gradually increased from 4 per cent to 9.3 per cent and then increased to 14.3 per cent. Compared to this, the increase in share of main workers engaged in other service sector jobs (include includes financial intermediaries, real estate activities and categories from other service sectors) has been sharper (increased from 15.2% to 17.2% and then to 21.6%) during the same period (Table: 19.1).

The educational status of the main workers and the distribution across different sectors indicate that workers with low educational status remain in some traditional jobs in primary sector while workers with higher educational qualification tend to concentrate in various jobs available in the service sector. Thus, majority of the illiterate workers are engaged in agriculture and allied jobs (nearly 58%), construction (11.8%) and manufacturing (10.5%) sectors. Compared to this, workers with educational status up to matriculation or secondary level have more presence in other sectors. It is interesting to note that only few workers with higher educational status engage in traditional agriculture allied sectors while their presence in various jobs available in service sector remain very high (Table: 19.2).

Table 19.2: Main Workers and Level of Education in Kerala - 2011 ***(Share %)***

Sectors	Total	Illiterate	Up to matric	Matric & above
Agriculture, Hunting, Forestry & Fishing	25.82	57.88	27.60	14.73
Mining and Quarrying	0.67	1.09	0.75	0.37
Manufacturing & Repairing	13.60	10.46	13.88	13.38
Electricity, Gas and Water Supply	0.59	0.40	0.51	0.87
Construction	14.28	11.77	15.54	11.20
Wholesale & Retail	11.85	5.69	11.62	13.75
Transport & Storage	8.95	2.52	9.40	8.91
Accommodation and food service activities	2.37	1.74	2.61	1.83
Information & Communication	1.42	0.12	1.10	2.60
Finance, Insurance; Real Estate; Professional, Scientific & Technical activities	3.27	0.32	2.58	5.81
Public Administration, Defence,				
Compulsory Social Security, Support Services etc.	4.58	0.90	3.65	7.97
Education, Human Health and Social Work activities	7.60	1.09	5.92	13.68
Other services (R-U)*	4.98	6.01	4.85	4.90
Total	100	100	100	100

*Arts, Entertainment and recreation; Other Service Activities; Activities of Households as Employers: Undifferentiated Goods and Services; Activities of Extra-Territorial

Source: Census of India

Within the service sector, electricity, gas and water supply; financial intermediation, real estate and business activities; and public administration, defence, etc. attract workers with higher educational status while other sectors like construction, retail and wholesale trade, hotels and restaurants, and transport, storage and communication etc may continue absorbing workers with basic levels of education. At the same time, this has led to situation where the educated youth refused to accept employment that they consider below their educational status or sought options outside the state. With preference for 'white-collar' jobs, especially in the organised sector, the educated work-force was not willing to engage in traditional farm jobs, despite being unemployed and the same led to shortage of workers in many areas (Mathew, 1999; Nair 1999). The nature of the state economy in terms of its dependency on service sector and stagnant commodity producing sectors, sustainability of employment growth is very limited.

4. Dynamics of Migration

Kerala experiences high rate of unemployment; limited growth in productive sectors, nature of growth, preferences of educated work-force, segmentation of workers and informalisation etc. contribute towards high incidence of unemployment, especially among the educated youth. When the strategy of prioritising social sector and progress of mass-education was not complemented by growth, the productive sectors could not absorb the educated workforce. At the same time, the success in mass-education enabled Malayalee workers in seeking new options outside the state.

4.1. Emigration of Malayalee workers

The contemporary Indian migration has a 175-year old history. By and large, the West remains as the favourite destination for skilled migrants from India while the unskilled migrants from India are concentrated in West-Asia (Khadria, 2009 & 2011). Until 1947 Kerala had a negative net migration, the state attracted more migrants than those who left the state. Since independence, there are two[2] kinds of international migration; migration of workers with skills and professional expertise, mainly to the USA, Canada, UK and Australia as permanent migrants (since early 1950s) and migration of unskilled and semi-skilled workers to the oil-exporting countries of West Asia on temporary work contracts, following the oil price increase of 1973-74 and 1979 (Rajan & Kumar, 2010). In the decade between 1981 and 1991, 555,000 persons migrated from Kerala to places outside India in addition to 189,000 who migrated to other states within the country (Zachariah, Mathew & Rajan, 2002). The period between 1980 and 1983 represent the peak phase of labour migration from India (a total of 976.7 thousand workers) where more than 50 per cent of them were Keralites (Prakash, 1999:137-139). The steady increase in emigration and inflow of remittances helped Kerala in entering a high growth phase ('virtuous growth') by late 1980s and emerged as the driving force behind the state economy.

Table 19.3: Educational Qualification of the Emigrants- 2018 ***(Share in %)***

Education Level	Male	Female	Total
Less than primary	2.8	8.8	3.7
Primary to Secondary	13.6	10.7	13.1
Secondary to Higher Secondary	42.8	11.2	37.8
Degree and above	24.6	53.1	29.1
Others	14.4	7.1	13.2
Illiterate	0.2	0.85	0.32
Haven't Started Schooling	1.8	7.82	2.73
Total	100	100	100

Source: Kerala Migration Surveys

Scholars have shared concerns around the dependency of economy on foreign remittances, its sustainability and leakages in utilisation of remittances, rehabilitation of returned emigrants, changing policy regime in the destination countries, and other policy issues. The fact that Kerala managed to continue the large-scale emigration for many decades indicates that the state can continue this model as long as it manages to explore new opportunities and prepare its workforce to make use of them. Analysis of international migration trends in India is inhibited by the limited official data available. The available information on workers migrating legally with Emigration Check Required (ECR)[3] from top sending states in India suggests that states like UP and Bihar experience steady growth while Kerala experienced marginal decline in recent years (International Labour Organisation, 2018). The fact that nearly 30 per cent of the total migrants have educational qualification of degree and above indicate the higher educational status of Malayalee migrants and potentials in this direction indicates that compared to earlier estimates, only a small share of the migrants is illiterate and with primary level educational status (Table: 19.3).

Table 19.4: Destination of Emigrants - 2018 *(Share in %)*

Destination	Male	Female	Total
Gulf	92	74.2	89.2
Australia	0.6	2.5	0.9
Canada	0.6	1.4	0.7
Singapore	0.5	1.2	0.6
United Kingdom	1.2	5.1	1.8
America	1.5	5.6	2.2
Others	3.5	10	4.6
Total	100	100	100

Source: Kerala Migration Surveys

While comparing the earlier generation of Malayalee migrants, the present generation of migrants have higher educational qualification and the same will enable occupational mobility and explore the opportunities in new areas. Thus, many of the Malayalee migrants are able to explore the opportunities in countries other than Gulf-countries (Table: 19.4).

The recent trend of emigration indicates some shift in its nature and dynamics. The recent round of Kerala Migration Survey (KMS) 2018 indicates a negative growth in emigration, especially towards Gulf countries. Compared to the constant increase in emigration during 1998-2013 period (34.9%), 2018 round of KMS reports a negative growth (11.6%) with a reduction of 3 lakh emigrants during 2013-18 (one-tenth of 2013 emigrants). Though Kerala experienced steady growth in emigration, this shift (even a marginal fall) indicates the possible limits of Gulf-migration from Kerala. At the same time, growth in the inflow of remittances continues (Table: 19.5).

Table 19.5: Trend in Emigration and Inflow of Household Remittances in Kerala (1998-2018)

Year	Emigrants	Remittances *(Rs. Crores)*
1998	1361919	3530
2003	1838478	7965
2008	2193412	12511
2011	2280543	15129
2013	2400375	24374
2018	2121888	30717

Source: Kerala Migration Surveys

Along with the declining trend in emigration, Kerala experiences a steady growth in the number of returned migrants (1.29 returned migrants in 2018-nearly 60% of emigrants). The fall in emigration can be linked to multiple factors or cumulative effects of demographic advances; low/constant wage in Gulf-countries and low savings; increase in the wages in domestic economy; decline in price of oil since 2010; nationalisation policies; options available outside, and other reasons (Rajan & Zachariah, 2018). The fact that the recent decline in emigration did not reduce the growth of remittances can be linked to improvement in quality of human capital, greater occupational mobility, occupational status, remitting capacity, weakening value of rupee and other factors.

4.2. Labour migration into Kerala: Domestic migrant labour

At macro level, the evidence suggests that there has been a steady increase in the share of migrants. For instance, 33 million migrants stated economic reasons for migration in 2001 while it has increased to 51 million in 2011 census respectively (Government of India, 2017: 266). In the case of Kerala, the emigration of workers out of the state and unwillingness of educated work-force to work in traditional sectors resulted in shortage of workers in many traditional sectors and wage works in Kerala. There are many push and pull factors like growing unemployment and lack of livelihood opportunities in home state; rural indebtedness in home state; high wage rate in Kerala; preference of employers in Kerala for migrant labourers; positive role of social network of migrants; pressure of large size of the family; poor income from agriculture; recruitment agents as catalyst triggering the migration of workers into Kerala (John, 2015). These factors seem to make Kerala as one of the favourite destinations for Domestic Migrant Labour (DML) in the country or DML emerges as the key source of wage workers in the state.

The long-term trend in inflow of DML into Kerala indicates steady growth in recent years. In 2013, one of the estimates suggests that Kerala has a stock of 25 lakh DML (equivalent to around 20% of Kerala's workforce). Compared to the earlier generation of DMLs who came from the neighbouring states like Tamil Nadu and Karnataka (many of them came along with families), the current generation are mainly from northern and north-eastern states (99% men and 91% within age group of18-35 years). Earlier, their presence was limited to specific sectors where there was shortage of manual labour but gradually expanded to other sectors (Narayana, Venkiteswaran & Joseph, 2013). Now, they have presence in almost all sectors while construction sector (nearly 60%), absorb the major share. The district-wise distribution of inter-state migrant workers varies across the districts in Kerala; Ernakulam has the highest share followed by Wayanad and Kannur (respectively 17%, 13% and 11%) districts (Figure: 19.1).

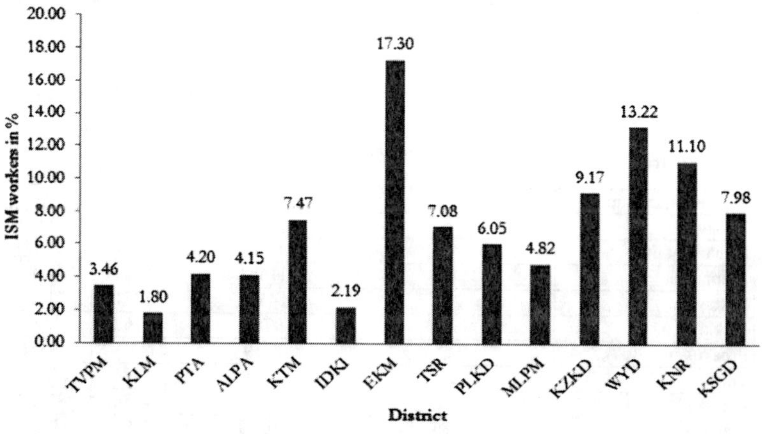

Source: Government of Kerala (2017:321).

Figure 19.1: Distribution of Inter-State Migrants in Kerala (Share in %)

Kerala has attracted migrant workers from different states; the current composition suggests that majority of them are from West Bengal, Odisha, Assam, Bihar, Jharkhand, Uttar Pradesh, Karnataka, Tamil Nadu and other

North-Eastern states. The details suggest that, it is possible to identify specific pockets and regions within these state where the majority of the workers belong to and the way the inter-state migration operates. Agents, contractors, relatives, friends, and other formal and informal ties strengthen these corridors and social network among the DMLs. Though Kerala state has initiated some specific policies including the introduction of live registry, social security measures and targeted service provisions, these workers continue to face challenges in conditions of work and living in different contexts (Government of Kerala, 2017; John, 2015; Narayana, Venkiteswaran & Joseph, 2013; Peter, & Narendran, 2017).

5. Employment and Migration in the Post-Disaster Context

Various agencies have tried to estimate the impact of disaster on Kerala and its society in terms of loss and damage. Given the nature of employment in the state, the available information is not adequate to provide any finite estimate on the loss of employment in different sectors of the state economy. The available figures show that disaster was found to have adversely affected both organised and un-organised sectors in the state. As a reflection of the differences in the distribution of these establishments and nature of disaster experienced across the districts, the loss of man-days and wage loss also vary across the districts (Table: 19.6). For instance, districts like Alappuzha and Ernakulam have large number of establishments and workers in both sectors and were severely affected by the disaster.

Table 19.6: Man-Days Lost in Various Establishments Due to Flood - 2018

District	No. of Establishment in organised sector	No.of Establishment in unorganised sector	No. of workers	Man-days Lost
Thiruvananthapuram	0	0	0	0
Kollam	13	21	270	15930
Pathanamthitta	393	6095	11371	125081
Alappuzha	19804	11683	69311	1043802
Kottayam	28	14689	17719	70876
Idukki	61	9786	57678	183223
Ernakulam	729	8439	112746	1195560
Thrissur	7	5645	19247	173396
Palakkad	27	947	12340	81141
Malappuram	2610	20290	47900	335300
Kozhikode	10	1744	21900	145674
Wayanad	4	394	1414	9798
Kannur	13	51	163	418
Kasaragod	0	0	0	0
Kerala	23699	79784	372059	3380199

Source: Labour Commissioner, Kerala

Similarly, the available information on person-days lost for workers registered under different welfare fund boards, fisheries and plantation sectors also indicate huge loss. In many cases, the damage to factories, industrial units, factories, working tools, raw materials, storage facilities and offices worsen the intensity and duration of disturbance and loss of livelihoods. In the case of fisheries and plantation sectors, the vulnerability caused by severe monsoon events disturbed their livelihood for several days. The incidence of major land slides in the hilly districts like Idukki, Wayanad, and Palakkad where the plantation activities are concentrated, thousands of plantation workers lost their jobs for many months. As per the records available with the Department of Labour, 35107 plantation workers (out of the 50882 workers in Kerala) were directly affected by the disaster. While considering the nature and duration of distress in the plantation sector, the government has introduced some special provisions for the affected workers.

The PDNA estimates the total damage and loss in the area of employment and livelihoods at Rs. 10,358 crore. The total damage (around Rs. 881 crore) covers tourism (Rs. 509 crore), MSME (Rs. 359 crore), and Kudumbashree and coir (Rs. 13 crore). The total loss was estimated around Rs. 9477 crore covering wage loss (Rs 7301 crore) and inventory loss (around Rs. 2176 crore). Along with this, total person days of employment lost (works out to 1097 lakh) and wage loss to the workers (estimated at around Rs. 7301 crore) also need to be considered (PDNA, 2018). Though different assessments followed different approaches and vary in the estimated cost of recovery, they emphasize the importance of restoring the employment and livelihoods as the primary step in the process of re-building Kerala.

Overall, the disaster weakened the productive sectors and economy which adversely affect the employment and livelihood options available. At the same time, the recovery strategies should consider the specific requirement and address the differential impact (if any) on different sectors and regions. For instance, addressing the vulnerability in unorganised sectors, traditional industries, plantation, and workers representing marginalised groups etc. may be prioritised in the recovery process. Similarly, modification of flagship programmes like Mahatma Gandhi National Employment Guarantee Scheme and other livelihood programmes like Ambedkar Employment Guarantee Scheme could be revived and modified in an effective manner. In the process of employment generation, measures should be taken to ensure its nature being sustainable and eco-friendly.

In the case of emigration, the evidence confirm the nature of Kerala's dependency on emigration and remittances. The KMS 2018 estimates show 2.1 million emigrants from Kerala across the world in 2018; and total household remittances at Rs. 85092 crore (Rajan, Zachariah, 2018). The fact that 40 per cent of the remittances received during the last 20 years in the state are used for housing and construction purpose. The nature of disaster and the damage it caused on land, housing and other valuables has pushed many people into distress and the same may trigger new wave of migration from the state. Along with the long history of successful emigration from the state, the nature of reconstruction, delay in process of restoring the pre-disaster livelihood options and limited growth of job in the domestic economy etc. may attract fresh emigration or emigration of returned migrants, at least in the short-run. Given the unfavourable demographic dividend, Kerala may not be able to sustain its dominance in the Gulf-region anymore. Its dependency on emigration and remittances will be compromised unless the state manages to improve emigration among the skilled workers and tap the opportunities that emerge in the developed world and promote utilisation of remittances into productive sectors in the state economy (Rajan, Zachariah, 2018). While recognising the critical role of the Malayalee diaspora in mobilising resources and support in addressing the rescue, relief and rehabilitation process in the state, state can use this context to devise strategic policy in enabling smooth and successful emigration from the state.

Considering the stock of inter-state migrants in 2013 (with 6.8% annual net flow of migrant labour), Kerala is likely to have around 35 lakh such workers in 2018. The estimates indicate that two-thirds of the migrant workers lost their workdays during the floods (affecting around 22.7 lakh and results in wage loss of Rs. 2,033 crore). If the wage loss to migrant workers is taken into account (taking a loss of 13 work days between 16 August and 30 August 2018), the total wage loss can be estimated around Rs. 2032.72 crore while the person days lost works out to 296.01 lakh (PDNA, 2018).

During the disaster, most of them had to leave their living place and shift to safer places; some of them returned to their homes. The state government also made efforts to help many of them to go back to their homelands by requesting special trains scheduled towards states sending large number of migrants to Kerala. There are many popular pockets where migrant workers live in large numbers in some of the districts, but very few accounts on the way they managed to survive the disaster. In some places, the local people arranged exclusive and special relief camps for the migrant labourers and extended all kinds of supports. Given the social stigma and divide that exist between local and migrant population, the volunteerism emerged between them was very impressive. Though many of them who went back to their home-states started returning to Kerala, the stagnant nature of the domestic economy and lack of jobs in certain sectors especially in the construction field pushed many of them to seek other options.

Kerala offers the highest wage rate in the unorganised sectors and the same has been the most attractive force behind this trend; but the current disaster and stagnant nature of local economy may not be able to sustain this high wage rate or it may bring some of the local workers back to the manual jobs which migrant labourers engage. These factors may reduce the inflow of migrant workers into Kerala or their gradual replacement. The discussion with the labour commission office suggested that there is hardly any official estimate on the number of DMLs who returned from Kerala or remained after the disaster (even among those who have formally registered as DML in Kerala). Given the nature of state's dependency on the supply of inter-state migrants, state needs to address the issues they face in the post-disaster context and ensure their contribution in the development in Kerala.

Conclusion

Kerala's experience can be used as an example to understand how mass-education affects the process of development in a regional context. The development experience of the state during the last few decades highlights the role of labour migration and inflow of household remittances in shaping the nature and pattern of development in Kerala. The evidence and discussion indicate that mass-education has its effect on the high rate of educated unemployment in the state, concentration of educated work-force in service sector jobs and withdrawal of educated workforce from traditional jobs in primary sector. The paper particularly discusses its impact on the dynamics of migration in terms of emigration of workers out of Kerala and in-migration into Kerala (Domestic Migrant Labour) to meet the shortage of manual workers. At one level, Kerala experiences high rate of unemployment and send nearly 22 lakh workers as emigrants. On the other hand, Kerala's labour market accommodates nearly 35 lakh migrant workforce from other states.

By looking into the available information, the paper makes an attempt to explore the implications of the recent disaster on employment situation and the dynamics of migration in Kerala. The loss of livelihood in traditional agriculture, plantation, fisheries and some of the traditional/ small-scale/cottage industries etc pushed many families into vulnerability. Many lost their livelihoods and tools of work. Policy intervention is necessary to restore their livelyhoods. The stagnation and delay in recovering from disturbances caused by the disaster may limit the options available to the migrant workers. With the support of the Malayalee diasporas across the globe and the larger public, the state government can use this context to revisit the policies effectively in mobilising the resources and expertise required for rebuilding the post-disaster state in a sustainable manner.

Notes

1. Government order No. (P)No.05/2018/DMD dated Thiruvanan-thapuram, 29.09.2018
2. Some studies identified the third phase of migration (from mid- 1990s) which is again led by professionals mainly software engineers and managers to the developed countries like United States and United Kingdom (Usha, 2009).
3. The limited coverage of this information (mostly for workers who have to register for emigration clearance) may be noted. It covers only who have not passed Class 10 at least, are leaving via employment visas into specific sectors or are workers emigrating for the first time to the GCC region, Malaysia and a few other countries.

References

1. *Census of India*, (2001 & 2011): Office of the Registrar General and Census Commissioner, Government of India
2. Government of India, (2017): *"Economic Survey 2016-17"*, Published by Department of Economic Affairs Economic Division, Ministry of Finance, New Delhi
3. Government of Kerala, (2017): *"Economic Review 2017"*. Published by the State Planning Board, Thiruvananthapuram
4. Government of Kerala, (2018): "Additional Memorandum: Kerala Floods - 2018" (1st August to 30th August 2018). Submitted by State Relief Commissioner, Disaster Management (Additional Chief Secretary)
5. International Labour Organization, (2018): India Labour Migration Update 2018 https://www.google.co.in/search? q=ILO+India+ Labour+Migration+Update+ 2018&rlz=1C1CHBFenIN822IN822&oq=ILO+India+Labour+Migration+ Update+2018&aqs=chrome..69i57.2046j0j7&sourceid=chrome&ie=UTF- 8# (Last accessed on 15th October, 2018)

6. Jafar, K. (2018): *Education, Migration and Human Development: Kerala Experience',* Rawat Publications, Jaipur.
7. John, J. (2015): "Interregional migration from Eastern States to Kerala: A Socio Economic Analysis", *Kerala Development Society (KDS Delhi) Working Paper No: 4,* August
8. Kannan, K.P. (2005): "Kerala's turnaround in growth: Role of social development, remittances and reform", *Economic and Political Weekly*, 40 (6): 548–554.
9. Kannan, K.P. (2007): "From human development to economic growth: Kerala's turn around in growth powered by human development, remittances, and reform", In A. Vaidyanathan and K.L. Krishna (Eds.), *Institutions and Markets in India's Development: Essays for K.N. Raj* (pp. 319-44), Oxford University Press, New Delhi.
10. Khadria, B., (2009): "India-Migration report 2009: Past, present, and the future outlook. International Migration and Diaspora Studies project", *School of Social Sciences at Jawaharlal Nehru University*. Cambridge University Press India Pvt. Ltd., New Delhi.
11. Khadria, B., (2011): "Bridging the binaries of skilled and unskilled migration from India". In, S.I. Rajan, and M. Percot, (Eds.), *Dynamics of Indian Migration: Historical and Current Perspectives* (pp. 251–285), Routledge, New Delhi.
12. Mathew, E.T., (1999): "Educated unemployment in Kerala: incidence, causes and policy implications", In B.A. Prakash (Ed). *Kerala's Economic Development: Issues and Problems* (pp. 94-115), Sage Publications, New Delhi.
13. Nair, M.K.S., (1999): "Labour shortage in a labour surplus economy? A study of the rural labour market in Kerala". In M.A. Oommen (Ed.), *Rethinking Development: Kerala's Development Experience* (pp 247–268). Volume 2, Institute of Social Sciences, Concept Publishing Company, New Delhi.
14. Narayana, D., Venkiteswaran C.S. & Joseph, M. P. (2013): "Domestic Migrant Labour in Kerala". *Gulati Institute of Finance and Taxation*, report submitted to Labour and Rehabilitation Department, Government of Kerala, Thiruvananthapuram.
15. Pani, N., & Jafar, K., (2010): "Mass education-led growth and non-agrarian villages: Long-term results of the Kerala model", *Oxford Development Studies*, 38 (1): 25–42
16. PDNA, (2018): "Kerala-Post Disaster Needs Assessment: Floods and Landslides - August 2018", *Joint report* prepared by the United Nations, European Union and Government of Kerala
17. Peter, B. & Narendran, V. (2017): "God's Own Workforce: Unravelling Labour Migration to Kerala", *Centre for Migration and Inclusive Development,* Perumbavoor- Kerala
18. Prakash, B.A., (1999): "The economic impact of migration to the Gulf", In, B.A. Prakash. (Ed.), *Kerala's Economic Development: Issues and Problems* (pp. 134–149). Sage Publications, New Delhi.
19. Rajan S.I., & Kumar, P., (2010): "Historical overview of international migration", In S.I. Rajan (Ed.), *India Migration Report 2010: Governance and Labour Migration* (pp. 1-29). Routledge, New Delhi.

20. Rajan, S.I. & Zachariah, K.C. (2018): *Kerala Migration Survey 2018,* Centre for Development Studies, Thiruvananthapuram.

21. Usha, L.R., (2009): "Migration from Kerala to the Gulf countries: Pattern, problems and policies". In B.N. Ghosh and P.D. Namboodiri (Eds.), *The Economy of Kerala: Yesterday, Today and Tomorrow,* pp. 243-258, Serials Publications, New Delhi.

22. Zachariah, K.C., Mathew, E.T., & Rajan, S.I., (2003): *Dynamics of Migration in Kerala: Dimensions, Differentials and Consequences,* Orient Longman, New Delhi.

23. Zachariah, K.C., Mathew, E.T., & Rajan, S.I., (2002): "Migration patterns and their socio-economics", In K.C. Zachariah, K.P. Kannan and S.I Rajan (Eds.), *Kerala's Gulf Connection: CDS Studies on International Labour Migration from Kerala State in India,* pp.13-45, Centre for Development Studies, Thiruvananthapuram.

24. Zachariah, K.C., & Rajan, S.I., (2012): *Kerala's Gulf Connection, 1998-2011: Economic and Social Impact of Migration*. Orient BlackSwan, New Delhi

25. Zachariah, K.C., & Rajan, S.I., (2013): *Diasporas in Kerala's Development*. Daanish Publishers, New Delhi.

20

Role of Animal Husbandry: Dairy and Poultry Sectors in Rebuilding of Kerala

K. Anilkumar

The cattle population in the state had declined by a factor 38% from 2003 to 2012. But milk production increased from 24.20 Lakh Metric tons to 27.16 lakh metric tons. Though there is a reduction in milk production from 2013-14 to the level of 25.20 lakh metric tons in 2016-17, the productivity of dairy animals of the state is on the increase. The contribution of livestock in Gross State Value Added (GSVA) in agriculture sector is nearly 29% amounting to around ₹ 10000 crore.

The unprecedented monsoon showers and subsequent flood situation in Kerala has resulted in loss in the tune of 40000 crore to the state. The estimated loss in Animal Husbandry sector is 5163 cows, 541 buffaloes, 5166 calves, 1228 heifers, 6380 goats, 14 lakh poultry, 4.6 lakh ducks and 1053 pigs and 20000 quails. Complete destruction of around 4500 cattle sheds, 3000 hectars of fodder cultivation and losses of stored concentrate feeds and feed ingredients are also reported. The after effects of flood on health of animals due to diseases like pneumonia, mastitits, stress related incidences of hemoprotozoan, bacterial and viral diseases are also likely to cause loss to the farmers in coming days.

Role of the animal husbandry, dairy and poultry sectors in rebuilding of Kerala are defined in six major areas. The sector can act as provider of livelihood security for the persons involved in it. These sectors are also providing supplementary income for agriculture based activities and stabilizing the agriculture based enterprises. Animal husbandry and related activities are job oriented and creates job opportunities better than most other sectors in the state. The job positions are the backbone of the rural economy and need to be encouraged. Another role these sectors are capable of playing is as an activity of continuous income generation. The sale of milk and other output of the enterprise ensures more or less continuous income generation many a times even on daily basis. This is one of the factors that keep the farmers glued to the sector irrespective of its low profit or loss situation. This is also true to some extent for activities like poultry egg production, broiler production and goat rearing. For a state like Kerala with a sizeable population abroad, animal husbandry activities can attract investments. Most of the expatriates are living with nostalgic feeling of farming and many of them like to invest in the sector. As Kerala is declared as a state with organic farming as the mode of agriculture, any system of organic farming requires animal manure as the major source of fertilization. The closely knitted agriculture and animal husbandry sectors is a must for establishment of organic farming units of various agriculture crops. Moreover manure and urine of cows, goats and other species help to maintain soil fertility.

A fifteen point action plan is suggested for the future planning of role of animal husbandry, dairy and poultry sectors in the rebuilding of Kerala:

1. Crossbred cattle are the major milk producers of the state. There is vast damage to the sector in the flood. Many animals were lost and farmers are finding it difficult to sustain their livelihood. The alternative is to introduce cows of high genetic merit into the state. It is possible to get good germplasm of crossbred from the military farms. The whole sector can be revamped with these introduced animals. The animals supplied should be bought under a central herd scheme. The performance needs to be monitored through out the year. The data recorded needs to be analysed for formulation of future strategies. All these animals should be insured and breeding need be done using superior semen. The major advantage through this programme is establishment of largest herd of crossbred cattle monitored continuously. This in turn can become the hub population for future genomic approaches in selection and improvement of the breed.
2. The productivity of crossbred animals in the state is increasing and the rate of increase need be accelerated. Application and popularization of new scientific methodologies like complete feeding, molecular technologies for disease screening, microchip based identification of animals and application of technology enabled reproduction are suggested.

3. Genomic screening of the animals and application of molecular technologies in selection process are necessary to ensure precision selection methodology.
4. There is a deficit of feed and fodder in Kerala and this situation can be addressed by use of locally available agriculture byproducts and feed ingredients. Feed mixing units need be established at milk cooperative society level for ensuring balanced ration and feeding for livestock.
5. Another aspect to be emphasized is to ensure clean milk production. The pricing of milk should be based on milk quality.
6. Conservation and multiplication of indigenous livestock and poultry recourses is another aspect to be stressed in the rebuilding of Kerala. Conservation units and periodic improvement of the indigenous stock is envisaged in future.
7. Meat animal production is to be enhanced as the conditions are favorable for maintaining large animals, small ruminants and pigs as meat animals. Units like male buffalo calves, male crossbred calves, crossbred pigs and indigenous goats are to be promoted. The packages suggested by the experts need be fine tuned.
8. These farming units need to ensure proper sanitization and environmental controls ensuring clean, green and ethical practices.
9. Establishment of breeder stock and hatcheries for broiler poultry chicks and layer chicks. The cooperative sector of the state can play an important role in these activities.
10. Comprehensive animal health monitoring system at door step a round the clock need be established. The "night vet" facility need be extended to all the veterinary hospitals / dispensaries of the state.
11. Insurance coverage to all the animals of the state is another programme suggested .

One of the major challenges is absence of experts for preparation of workable project proposals and establishment of the projects. If this aspect is addressed it is possible to have quantum jump in related activities.

21

Rebuilding the Flood-hit Kerala Economy: Imperatives for the Sustainable Growth of Kerala's Tourism Sector

P.K. Manoj

1. Introduction

It is widely recognised in the literature that tourism offers excellent potential for employment generation and economic development and hence tourism is being promoted aggressively by nations, particularly the developing nations like India. It may be noted that at the global level tourism is ranked second highest in terms of income generation and is second only to the oil industry. Tourism has emerged as one of the largest and fastest growing industries in the whole world. The combined effect of enlarged leisure and better purchasing power makes tourism affordable to more people. Every country can play a decisive role in tourism and tourism industry is reckoned as very valuable, particularly for the development of global transportation and communication. Of late, this industry is receiving good attention as a foreign exchange earner and in most parts of the world it is aggressively promoted primarily for attracting more foreign exchange. The huge amounts spent by foreign tourists for accommodation, food, transport, recreation, etc. have got linkage effects resulting in overall economic development of the host countries.

In India, after the independence of the country in 1947, the focus of the government was on the development of sectors like agriculture, industry, infrastructure etc. Tourism sector in India was never considered as a sector with high growth potential and it had been growing at slow pace during the early years of the country's independence. Hence, Indian tourism could not exploit the vast tourism potential of the country during those years. But, since the last two decades or more, there has been an organised effort to aggressively promote tourism by the Government of India. Quite similar to the poor tourism performance of Indian tourism among comparable world nations, the relative performance of tourism sector of Kerala has been very poor vis-a-vis other states in the Indian union. In spite of the enviable tourism resources in Kerala the state is lagging far behind most of the other Indian states, notwithstanding the fact that those states have low tourism endowments.

India has been showing the best way to strategize sustainable tourism by way of encouraging the pursuit of tourism in a responsible manner. It is about sensitizing tourists on the need for conserving and maintaining the natural wealth and cultural heritage of a nation. In the above context, environment-friendly tourism models, particularly the Responsible Tourism (RT) Model, has assumed tremendous significance in states like Kerala in India. Of late, the Government of Kerala has been giving focused attention on RT, though allied models like ecotourism, rural tourism etc. are also being accorded high priority. RT seeks to measure the effects of tourism on the well being of the host community, their art and culture, byproducts, hereditary values and knowledge, as well as on local environment. RT tourism envisages minimum negative effect on the culture of the indigenous society, and on the environment. So, RT ensures environmental sustainability besides magnifying host culture. Though RT and its variants like ecotourism have immense growth potential in Kerala because of the huge demand from foreign tourists, these models are yet to pick up adequate growth momentum in the state.

The flood that has occurred in Kerala in August 2018 has adversely affected all industries in the state, particularly the tourism industry. So, the recent flood has given another dimension to the constraints and limitations of Kerala tourism. Suitable remedial measures need to be taken to bring the flood-hit tourism back to its normalcy.

2. Literature Review and Research Gap

The industry report by World Economic Forum (WEF) (2015) [29] 'The Travel and Tourism Competitiveness Report' has done a detailed study of the competitiveness of various countries of the world, in the travel and tourism front. Competitiveness of nations at the regional and global levels based on a number of well-defined parameters is dealt in the WEF report. As per WEF report, the global competitiveness of Indian tourism is only 52 as against 17 in respect of China. This indicates a huge performance gap between the two comparable (and neighbouring) nations of the developing world. At the regional (Asia Pacific region) level too, there is a huge gap in the competitiveness, as India is in the 12th position whereas China is ranked in the 6th position. WTTC (2015) [32] in its latest report, 'Economic Impact of Travel & Tourism 2015 - Annual Update' has made an elaborate account of the relative performance of various countries of the world in the tourism front, including region-wise performance and future projections. WTTC has reported that South Asia, led by India and the Middle East, is globally the fastest growing region in terms of the total contribution of Travel and Tourism to GDP. Accordingly, India is one among the bigger, fast growth markets along with China, Indonesia, South Korea and Turkey. Besides, WTTC has reported that South Asia will be the fastest growing sub-region for total Travel & Tourism GDP long-run growth (7.0 percent) as India outpaces China. Empirical studies on ecotourism in the Indian context, particularly in the context of Kerala state in the Indian union are rather rare. Some relevant studies are briefly discussed here. Kumar, Yathish (2007) [12] has noted that the aim of tourism is to improve the quality of life of people, provide a good experience to the tourists, and maintain the environmental quality which is vital for both the tourists and the local populace. A macro level study done in the Kerala context by Manoj P K (2008) [16], 'Sustainable Tourism in India: A Study from a Global Perspective with Focus on Tourism Prospects of Kerala' points out the vast growth prospects of sustainable tourism in Kerala from a global perspective. The author has suggested some strategies for the faster growth of sustainable tourism with special reference to Kerala. In another Kerala-based study, Oommen M. A. (2008) [23] warned about the danger of overlooking the environmental sustainability in development activities in the state. While referring to the growing environmental issues in Kerala from the perspective of the state's economic development model viz. 'Kerala model of development' the author has pointed out in detail the environmental and ecological issues that the state faces at present; the current scenario being critically referred to as one of 'Ecological Overkill' in his paper. The day by day worsening situation of the natural environment in Kerala and its adverse impact on the long-term sustainability of tourism has been noted by many researchers, pointing out the need for environment-friendly (nature-friendly) tourism models like ecotourism and its variants like rural tourism, responsible tourism etc. Notably, the macro level study by Manoj P K (2009) [17], 'Environment Friendly Tourism for Sustainable Economic Development in India', for instance, underscores the cardinal significance of tourism in India for its rapid economic development; and at the same time point out the need for promoting environment-friendly (and hence sustainable) tourism models. The author, thus, makes a few strategies for development of environment-friendly tourism in India. In his book on 'Ecotourism Development Management', Singh, Sarvjeet (2009) [25] has stated that ecotourism is entirely a new approach in tourism and it provides opportunities for visitors to experience powerful displays of nature and culture and to learn about the importance of biodiversity, conservation and local cultures. It focuses on volunteering, personal growth and finding innovative means to live on the earth. It involves travels towards locations wherein flora and fauna, cultural heritage etc. are the main attractions. It encourages the active participation by the local population in the conservation and education dimensions of tourism development process.

In a research article by Manoj P K (2010) [18], 'Tourism in Kerala: A Study of the Imperatives and Impediments with Focus on Ecotourism', the tourism sector in Kerala state in India is analysed in detail using the SWOT model. Accordingly, strategies are suggested for the faster and sustainable development of tourism in Kerala. Sudheer, B (2015) [26] in his research report submitted to University Grants Commission (UGC), New Delhi titled 'Economic and Cultural Impact of Responsible Tourism Initiative in Kerala - A Case Study of Kumarakom Panchayath' has highlighted the need for alternative and innovative practices like Responsible Tourism (RT) to minimise the negative effects of tourism on environment and to make it sustainable in the long-term. With respect to RT at Kumarakom in Kerala, the positive effects like employment to the local people, empowerment of women through RT-related activities (like, providing vegetables, fish, meat etc. often procured locally), positive linkage effects on the locality (like, earnings arising from the purchases made by tourists) etc. have been pointed out.

Though there are many studies on tourism done in the Indian context, including a few studies in the Kerala context, empirical studies that focus on the impact of flood on Kerala tourism are virtually nil. So, this study seeks to bridge the above research gap.

3. Relevance and Significance

The concept of inclusive growth is of utmost significance in India, and in Kerala which has got huge tourism potential. Though tourism is being aggressively promoted by the Government of Kerala, particularly the Responsible Tourism (RT) model (including its variants like ecotourism, rural tourism etc.), tourism is yet to pick up growth momentum in the state. The flood havoc that took place in Kerala in August 2018 has adversely affected all industries, particularly the tourism sector. In the above context, it is relevant to make a focused study of the impact of flood on Kerala tourism with a view to suggest remedial measures, notwithstanding the fact that Kerala tourism is fast approaching normalcy after the flood havoc. So, this paper studies the impact of flood on Kerala tourism and accordingly suggests suitable remedial measures.

4. Objectives

(i) To study the prospects and problems of Kerala tourism, in terms of foreign exchange earnings, foreign tourist arrivals etc., with a focus on responsible tourism (RT);

(ii) To study the impact of flood havoc on Kerala tourism and its current status; and

(iii) To suggest strategies for the growth of Kerala tourism and hence the Kerala economy.

5. Methodology of the Study

This study is descriptive-analytical and exploratory in nature. Both primary and secondary data are used for this study. Primary data are collected by means of sample survey using an Interview Schedule. Secondary data are collected from the publications of United Nations World Tourism Organization (UNWTO), World Travel and Tourism Council (WTTC), Kerala Tourism Development Corporation (KTDC), Kerala State Planning Board, tourism statistics of the tourism departments of Governments of Kerala and India, reports on tourism industry by agencies like FICCI, IBEF etc. Popular statistical tools are used for data analysis.

6. Prospects and Problems of Tourism in Kerala

In fact, way back in 2000 Kerala had drafted a comprehensive tourism policy viz. *Vision 2020* that had given due consideration to the sustainability aspects of tourism growth along with its economic and feasibility aspects. This policy sought to make tourism one of the core competencies of Kerala and to regulate the tourism initiatives properly taking into account the broader and long-term interests of Kerala, like, environmental conservation, protection of heritage and culture, development of tourism infrastructure, quality standards etc. In short, it aims at using a sustainable and integrated approach to tourism promotion in Kerala. The performance of Kerala tourism has been impressive, both in absolute and relative terms. There has been an increasing trend in respect of foreign tourist arrivals to Kerala over the years and also in the share of Kerala in the tourism revenue for the whole of India.

Despite the good growth prospects for Kerala tourism, there is scope for further improvement in view of its enviable tourism potential as the "God's own country". It is noted that there are a few areas wherein Kerala has a comparative edge over the rest of India and also other countries of the world. Its unique socio-economic and geographic profile has added significantly towards making it one of the sought-after tourist destinations of the world. The lengthy coastal belt, peculiar geographical location, lusty backwaters, a large number of beaches, serene hillocks, moderate climate throughout the year, highly literate populace etc. are some of the factors. It is noted that there has been a growing trend in respect of foreign tourists and domestic tourists to Kerala over the 18 years' period, 2000 to 2017 (Table I).

Table I: Trends in Domestic and Foreign Tourist Arrivals into Kerala (2000-2017)

Year	Domestic Tourists	Percentage Change	Foreign Tourists	Percentage Change
2000	5013221	2.6	209533	3.8
2001	5239692	4.5	208830	-0.5
2002	5568256	6.3	232564	11.3
2003	5871228	5.4	294621	26.7
2004	5972182	1.7	345546	17.3
2005	5946423	-4.3	346499	.27
2006	6271724	5.47	428534	23.7
2007	6642941	5.92	515808	20.37
2008	7591250	14.28	598928	16.11
2009	7913537	4.25	557258	-6.96
2010	8595075	8.61	659265	18.31
2011	9381455	9.15	732985	11.18
2012	10076854	7.41	793696	8.28
2013	10857811	7.75	858143	8.12
2014	11695411	7.71	923366	7.60
2015	12465571	6.59	977479	5.86
2016	13172535	5.67	1038419	6.23
2017	14673520	11.39	1091870	5.15

Source: Govt. of Kerala (2018), Economic Survey; and Tourism statistics (Kerala Tourism).

The Department of Tourism (DoT) of the state has formulated a three-pronged strategy to maintain and further improve the superior position of Kerala in the global tourism map. Accordingly, 3 broad strategies viz. (i) Product Innovation, (ii) Innovative Marketing, and (iii) Strategic alliance with the Government of Rajasthan. Besides, there are efforts in the direction of obtaining the benefits of Special Economic Zones (SEZs) for tourism, so that SEZs for tourism could be established meaningfully. The recently drawn up 'Kerala Tourism Policy 2017' has given a further boost to RT initiatives in Kerala whereby Responsible Tourism Mission (RT Mission) will be implemented to promote RT in the state in a systematic manner.

7. Impact of Flood on Kerala Tourism and the Present Status of Kerala Tourism

Based on the discussions with all the stakeholders, the following are the adverse effects of the recent devastating flood that occurred in Kerala in the mid-August 2018:

(i) Transportation was disturbed during one week and three weeks partially;

(ii) There was lack of proper communication facilities for nearly two weeks;

(iii) There was cancellation of registrations to the tune of 60 percent in August 2018, but the situation gradually improved to 20 percent or less by the end of October 2018.

(iv) Image of the tourism spots was affected after the flood, but later it gradually improved.

In spite of the above temporary issues, because of the pro-active steps taken by the Government, particularly the Tourism Department, the tourism industry in Kerala has been gradually attaining normalcy. The adverse impact of the floods is vanishing and more tourists are expected in the future. The macro level (Kerala state as a whole) picture on the impact of the unprecedented flood on Kerala tourism are as follows:

- Foreign Tourist Arrivals (FTAs) decreased by 13.55 percent during the third quarter viz. July-Sept. 2018 compared to the same period last year i.e. July-Sept. 2017. There was a positive growth of 12.13 percent in FTAs in the first quarter, viz. Jan.-March 2018. But, a negative growth of -4.6 percent was registered in the second quarter viz. April-June 2018; because of the alleged rape and murder of a Latvian woman near Kovalam in Kerala and also the Nipah virus outbreak in Kozhikode in May 2018. This negative growth further widened to 13.55 percent in the third quarter viz, July-Sept. 2018, as already noted above.

- Domestic Tourist Arrivals (DTAs) decreased by 3.48 percent during July-Sept. 2018 whereas there was a high increase of 15.93 percent during May-June 2018. This is for the first time that the state witnessed a negative growth rate in DTAs, as the normal trend in Kerala used to be that of 5-6 percent annual overall growth rate, and double digit growth during special occasions. The four worst-hit districts because of the devastating flood were Idukki, Wayanad, Alappuzha, and Thrissur.
- Flight bookings to Kochi (via. Cochin International Airport Limited (CIAL) almost halved to 53 percent in August and Sept. 2018 - a direct impact of the floods. In October too, the flight bookings was 17 percent lower than that of last year, i.e. Oct. 2017.
- The often attributed weakness of 'lack of hygiene' became worse after the flood, because the flood further deteriorated the cleanliness in the tourism spots. Another persisting complaint is that of exploitation of tourists, particularly the foreign tourists.

Flood has caused irreparable damage to Kerala economy, particularly Kerala's tourism sector.

In spite of the adverse impact of flood on Kerala tourism, there are indications of a revival of the sector. Some indications are as follows;

- Foreign tourists arrived in large numbers, over 900 tourists - all from United Kingdom, in three charter flights in the month of December 2018.
- Large number of tourists started flowing to Munnar in December 2018, increasing the occupancy levels in the hill stations and home stays in that tourism spot.
- Gradually improving status of health and wellness tourism in Kerala, particularly the renowned Kerala brand Ayurveda tourism. Other types of medical tourism (like, modern medicine) is also picking up in the state, though initially affected by the flood.

8. Suggestions Based on the Findings of the Study

- Kerala has already formed the Responsible Tourism Mission

(RTM) in 2007. Responsible Tourism (RT) deserves special attention in Kerala. RT ensures balanced development, particularly the development of the local community. RT models like the one in Kumarakom in Kottayam is a role model for the whole nation. The Kumarakom model needs to replicated in other destinations too, but considering the local conditions.

- Events like Kerala Travel Mart (KTM) need to be arranged periodically so as to promote tourism in Kerala in a greater way.
- Remedial measures are required to counter the negative impact of tourism, particularly mass tourism. The environmental degradation because of plastic littering, imbalance in bio-diversity, pollution of water bodies and streams etc., as observed in many ecotourism destinations (like, Thenmala, in Kollam district) needs to be controlled. Mass tourism should give way to 'class tourism' so as to ensure the sustainability and environmental purity. Control is required on the number of tourists in busy destinations.
- Less known destinations have to be promoted so as to reduce the intensity of tourists in well known destinations. This is because of the fact that in many reputed tourism spots the tourist arrivals are more than the carrying capacity of the destination, and this in turn affects the long term sustainability of the respective tourism spots.
- Diverse tourism models, including newer and less common ones like cruise tourism, adventure tourism etc. can be successfully implemented in Kerala, given the unique geographic advantages of the state, like the vast coastal areas, serene hillocks etc.
- The problems of lack of hygiene and exploitation of tourists are two worst problems. Efforts are required to get rid of these evils from the tourism products on the offer.
- Many tourists are interested in healthcare tourism, and some are interested in studying about traditional products like those based on Ayurveda (Ayurveda tourism). Hence, the government and other stake holders should provide facilities to encourage such tourists, including those interested in such studies. This would enhance the image of the sector.
- Concerted efforts towards aggressive promotion of Kerala tourism is necessary. Given the immense potential of information and communication technology (ICT), the same has to be used not

only for tourism promotion but also as a vital element that is embedded in all tourism products and services. ICT is an enabler of superior tourism products and services and also enhanced productivity, and hence ICT adoption needs to be nurtured.

- Special thrust is required on Human Resource Development (HRD) in the tourism sector, particularly with the help of modern ICT-based tools. Only highly skilled and techno-savvy professionals can provide the kind of service that is expected by the growingly discerning tourists from advanced nations of the world, like the US or UK. Extensive use of ICT in all facets of the delivery of tourism products and services is an imperative for the growth of the sector in Kerala, as is the case elsewhere in the world. Special training should be provided to tourist guides on ICT-based tools and applications.

9. Conclusion

In spite of the adverse impact of flood on Kerala tourism, it is fast picking up in the state. By way of concerted action Kerala can definitely attain complete normalcy in tourism sector. The indications have so far been positive and let us hope that Kerala tourism will scale new heights.

References

1. Ajith Kumar M.K. (1998): "Psychographic & Demographic Profiles of Foreign Tourists Visiting Kerala with special reference to Spending Patterns", *Unpublished Ph.D. Thesis*, Cochin University of Science and Technology (CUSAT), Kochi, Kerala.
2. Ballabh, Aanandh (2005): *Fundamentals of Travel and Tourism*, Akansha Publishing House
3. Brundtland G.H. (1987): *The Brundtland Report, World Commission on Environment and Development*, Oxford University Press.
4. Bhatia, A.K. (1996): *Tourism Development: Principles and Practices*, Sterling Publishers Pvt. Limited, New Delhi - 110 020 (India).
5. Bijender K. Punia (1994): *Tourism Management - Problem and Prospects,* pp. 169, Ashish Publishing House, New Delhi
6. Chatak, G.R. (2007): *Sustainable Tourism Codes and Guidelines*, Cyber-Tech Publications, Daryaganj, New Delhi, (First Edition)
7. Clark, J.R. (Ed.) (1991): *Carrying Capacity: A Status Report on Marine and Coastal Parks and Reserves,* University of Miami/US National Park Service, Washington, DC.
8. Clifford J (1998): *The Predicament of Culture: 20th Century Ethnography, Literature and Art, Cambridge,* MA: Harward University Press.

9. Brandon, K. (1996): Ecotourism and Conservation: A Review of Key Issues. Washington, DC: The World Bank.

10. Government of Kerala, *Economic Review,* State Planning Board, Thiruvananthapuram, Kerala - for the years 2008 to 2017. [Available online at https://spb.kerala.gov.in].

11. Kabia, Sunil K. (2005): *Tourism and Environment*, Mohit Publications, New Delhi, (First Edition).

12. Kumar, Yathish (2007): "Tourism Sector and Sustainable Development", *Southern Economist,* Vol. 45, No. 9.

13. Lane, B. (1991): "Sustainable Tourism, A new culture for the Interpreter", *Interpretation Journal*, Vol. 49.

14. Lindberg, K., and J. Enriquez (1994): *An Analysis of Ecotourism's Contribution to Conservation and Development in Belize*, Vol. 2. Washington: World Wildlife Fund.

15. McNeeely, J.A., J. Thorsell., and H. Ceballos-Lascurain (1992): *Guidelines: Development of National Parks and Protected Areas for Tourism*, WTO/UNEP/IUCN, Madrid.

16. Manoj P. K. (2008): "Sustainable Tourism in India: A Study from a Global Perspective with Focus on Tourism Prospects of Kerala", *Proceedings of Second International Conference on Responsible Tourism in Destinations*, 21-24 March, 2008, p.7. Also available online in the [Official website: http://www.artyforum.info/documents/MicrosoftWord-MANOJ-Poster.pdf]

17. Manoj, P. K. (2009): "Environment Friendly Tourism for Sustainable Economic Development in India", *International Journal of Commerce and Business Management,* pp.139-147.

18. Manoj P K (2010): "Tourism in Kerala: A Study of the Imperatives and Impediments with Focus on Ecotourism", *Saaransh RKG Journal of Management,* Vol. 1, No. 2, Jan. 2010, pp.78-82.

19. Manoj P K (2015): "Prospects of Eco tourism in Kerala: Evidence from Kumarakom in Kottayam District" *International Journal of Research in Management and Social Science,* Vol. 3, Issue 1 (II), Jan. – March 2015, pp. 54-59.

20. Manoj P K (2015): "Employment Generation from Rural Tourism: A Field Study of the Local Community at Kumbalangi, Kerala", *International Journal of Applied Services Marketing Perspectives (IJASMP),* Vol. 4, No. 4, Oct.-Dec. 2015, pp.1880-1888.

21. Manoj P K (2016): "Impact of Rural Tourism on the Environment and Society: Evidence from Kumbalangi in Kerala, India", *International Journal of Advance Research in Computer Science and Management Studies (IJARCSMS)*, Vol. 4, Issue 2, Feb. 2016, pp.148-159.

22. Manoj P K (2017): "Segmentation Strategy for Promotion of Ecotourism Products: Evidence from Thenmala Ecotourism", *South Asian Journal of Socio-Political Studies (SAJOSPS)*, Vol.18, No.1, July-Dec. 2017, pp.112-119.

23. Oommen, M.A., (2008): "Reforms and the Kerala Model", *Economic & Political Weekly*, Vol. XLIII, No.2, Jan. 12-18, 2008, pp. 22-25.

24. Singh, Ratandeep (2008): *Tourism Marketing Principles, Policies and Strategies*, Kanishka Publishers & Distributors, Darya Ganj, Delhi - 110 002 (India).
25. Singh, Sarvjeet (2009): *Ecotourism Development Management,* A.K. Publications, New Delhi
26. Sudheer, B. (2015): "Economic and Cultural Impact of Responsible Tourism Initiative in Kerala - A Case Study of Kumarakom Panchayath", *Research Report,* Research Sponsored by the University Grants Commission (UGC), New Delhi. (www.sncollegecherthala.in).
27. Verma, J. (2007): *Dynamics of Indian Tourism Issues & Challenges*, Cyber-Tech Publishers, Daryaganj, New Delhi, (First Edition).
28. ATREE (2006): *White Paper on Ecotourism Policy* (Draft), April, pp 4-5.
29. WEF (World Economic Forum) (2015): *The Travel & Tourism Competitiveness Report 2015,* 91-93 route de la Capite, CH-1223 Cologny, Geneva, Switzerland.
30. United Nations World Tourism Organization (UNWTO) (2015): *UNWTO Tourism Highlights 2015 Edition*, Calle Capitán Haya, 42 28020 Madrid.
31. Weaver, D.B. (2001): *Ecotourism,* John Wiley & Sons: Milton
32. World Travel & Tourism Council (WTTC) (2015): *Economic Impact of Travel & Tourism 2015 Annual Update*, Harlequin Building, 65 Southwark Street, London, SE1 0HR, United Kingdom.
33. Government of Kerala (2014): *Kerala Tourism Statistics 2014*, Research & Statistics Division, Dept. of Tourism, Thiruvananthapuram.
34. Government of India (2014): *India Tourism Statistics at a Glance 2014*, Dept. of Tourism, New Delhi.
35. Official website of World Tourism Organisation (WTO), www.unwto.org
36. Official website of World Travel and Tourism Council (WTTC), www.wttc.org
37. Official website of Kerala Tourism (Government of Kerala), www.keralatourism.org
38. Official website of Incredible India (Government of India).

Part - III

Learning Lessons and Experiences

22

Of Flows and Flawed Frames: What Can We Learn (and Unlearn) from the International Experience on Floods?

Nimmi Kurian

The Analytical Cul-de-sac

The paper will engage with the extensive international literature on flood risk governance so as to identify both the opportunities and barriers towards more inclusive models of disaster management. Engaging with this literature can also be a useful first step towards designing norms of benefit sharing, negotiating trade-offs, and allocating risks and burdens that have a direct bearing for the life chances of local communities. How do we understand many of these intersections in India's immediate and extended neighbourhood where resource degradation is creating and reinforcing interlocking webs of environmental and socio-economic vulnerability? Parts of Asia are already severely water stressed and are likely to witness an increase in the frequency and intensity of extreme events such as floods and droughts. The Ganges, Brahmaputra and Meghna basins are among those endemically flood-affected regions of the world and account for 61 per cent of the country's land surface and up to 85 percent of its flood losses (Ghani: 2002). Across South Asia, climate change is expected to bring an assorted set of challenges ranging from recurrent floods, droughts, declining water tables and a non-linear pattern in monsoons adversely affecting lives and livelihoods (IPCC, 2007: 476). The projected sea level rise could result in a rise of 40 cm, with an estimated 94 million people estimated to be affected due to flooding in coastal areas (IPCC, 2007: 484).[1] Agrarian-based economies are particularly prone to the vulnerabilities that climate change brings in its wake. For instance, loss of biodiversity is already adversely affecting the livelihoods of forest communities living on the India-Myanmar and India-Bangladesh borderlands. For an estimated 4.5 million people living in the low-lying Indian Sundarbans area, any change in sea level translates into a direct threat to their lives and livelihoods. Studies suggest the probability of a sea level rise of 50 cm by 2050 that could submerge large parts of the Sundarbans delta (WWF, Sundarbans, *Future Imperfect*). One of the greatest impacts of climate change in this area comes in the form of displacement. In West Bengal, 7,000 people have so far reportedly been displaced from the Sundarbans region due to sinking of the islands, and by 2030 this number could well increase to 70,000 affected by flooding (Hazra, 2006).

To begin with, one needs to take a harder look at three interrelated themes that are central to the discourse on floods, namely the notions of opportunity, learning and disaster. Firstly, what sort of an 'opportunity' does the crisis represent and for whom is it an opportunity? In this context, it is useful to recall Naomi Klein's powerful critique of the kind of choices that tend to be made after disasters. These, as she points out, have tended to invariably favour neoliberal, free-market interventions. The 'state of shock' as Klein argues, has offered fertile ground for American capitalism to thrive and profiteer. Recall for instance, how in the aftermath of the U.S Civil War of 1861-1865, the quintessential outsider from the North relocated to the war-torn South in search of business opportunities in the rubble of war-devastated economies. They came to be by and large viewed in the public perception of Southerners as opportunistic carpet baggers.[2]

Further, the idea of 'learning' also needs to be problematised to debate what is the nature of the learning that is being referred to. For instance, the idea of 'New Kerala' that is being projected appears at first glance to be a fairly straight forward feel good narrative. It talks of reconstruction, of restoration and of new beginnings but there is also a hint at forgetting and erasing the past that may not be entirely innocent. Implicit in it is the danger of a wilful erasure of painful memories and thereby of the opportunity to learn (as well as unlearn) from the disaster. A troubling case in point is the aftermath of the Gujarat earthquake of 2001 and the nature of reconstruction models that the state chose to adopt. These, as Edward Simpson reminds us in *The Political Biography of an Earthquake: Aftermath and Amnesia*, were designed precisely to wipe out such difficult memories and 'deliver a brand new state'. These raise the question that Kerala needs to ask of itself: can it afford to forget? In marked contrast to

Gujarat is the manner in which Chile chose to respond to disasters. These could, in many ways offer clear parallels for Kerala given that Chile is a tiny strip of land as well as predominantly coastal like Kerala. Unlike Gujarat, Chile consciously chose to remember and not to erase the painful memory of the disaster. In many ways, it refused to forget by investing in daily reminders such as evacuation sirens and effectively using popular culture.

The idea of 'disaster' needs to be similarly problematised to understand how it is framed in the mainstream research and policy imagination. A few illustrative examples can give a sense of the manner in which India has dealt with these framing issues and the price it has paid for the flawed framing of the issue. The first instance of flawed framing is the 8.6 magnitude Assam earthquake and floods of 1950 that was as Angus McLeod Gunn points out, not 'an Indian earthquake' but had its epicentre in Rima in Eastern Tibet (Gunn 2008: 417). But while the epicenter of the earthquake lay in southeastern Tibet, it was the Brahmaputra Valley that bore the most extensive damage. The earthquake blocked several tributaries of the Brahmaputra and created a 'wall of water' measuring thirty feet high destroying several villages in its wake. The Tibetan "water bank" is in every sense Asia's water bank and the environmental sustainability of Tibet means the environmental sustainability of much of Asia. It constitutes the headwaters of many of Asia's mighty rivers including the Brahmaputra, Mekong and the Yangtze that flow into some of the most populous regions of South and Southeast Asia (Kurian 2004). Studies by Chinese scientists are also pointing to the possibility of a high content of heavy metals in the stream sediments and tailings in Tibet's Gyama Valley that could pose a potential threat to downstream water users (Huang 2010). Added to these are concerns that the fragile ecosystem of the Tibet-Qinghai plateau is showing signs of stress as it struggles to cope with the furious pace of economic activity. The 'pillar industries' of mining and timber processing have fed the rapid industrialisation of Tibet, bringing in its wake assorted problems of deforestation, soil erosion, landslides, floods, acid rain and pollution especially of its water systems. These are creating ecological imbalances in the form of rising temperatures, retreating glaciers and droughts caused by indifferent rainfall (Kurian 2018). These are sobering reminders of the need to bring scale into any conversation on disaster risk management. Scale, particularly transboundary scale, remains an understudied dimension in disaster preparedness and response. Boundary-spanning crises such as floods or earthquakes require new scales of governance within cross-border regions.

A second instance of flawed framing is the phenomenon of black Brahmaputra in December 2017, stoking furious speculation that Chinese upstream interventions were causing the colour of the downstream waters to turn black. The speculations also included the allegation that China was constructing a 600-km long tunnel in Yunnan with the purpose of diverting the waters of the Yarlung Tsangpo to the Taklamakan desert in Xinjiang (*Times of India* 2017). While there was considerable downstream impact, the assumption that China was responsible for it was a flawed correlation in many ways. A recent study dismissed this correlation and established that it was in fact an earthquake in Tibet on 17 December that had resulted in the muddying of the waters of the Brahmaputra.

A third instance of flawed framing is the Kosi floods of 2008, which stands as a disturbing metaphor for much that is wrong with the mainstream Indian discourse on water. The breach in the Kosi barrage in Nepal on 18th August resulted in a humanitarian disaster across both Nepal and India with nearly 50,000 in Nepal and 3 million people in the Indian state of Bihar displaced. Nearly 247,000 acres of farmland was also estimated to have been destroyed in the process. Despite this, the mainstream thinking on water has remained in a state of denial despite mounting evidence to the contrary. Take for instance the official definition of the Kosi floods as a 'natural' disaster. The belief in embankments as a flood control measure stands largely uncontested. Despite its stated capacity of 9.5 lakh cusecs of water, the Kosi embankment in Nepal breached with just 1.44 lakh cusecs due to the heavy accumulation of silt deposits. Bihar for instance, has 3600 km of embankments but what is troubling is that the area considered flood-prone nearly tripled from 250,000 hectares in 1952 to more than 688,000 hectares in 1994. With the 368,000 ha of the 'flood-protected land' that was flooded in the breach of the Kosi River in 2008, the flood-prone area of the state rose to 724,800 ha (Mishra 2011). Kosi's annual siltation deposits of over 80 million tons have also eroded the East Bank canal's irrigation potential to just 7 per cent.

Another instance is the 2000 flash floods in Himachal Pradesh caused due to the formation of an artificial lake in Tibet. Such crises underscore the need for early warning systems and to perfect coordination mechanisms including visits to sites as well as putting up permanent monitoring stations to enable quick transmission of information. This fact was brought home in a tragic manner during the flash floods caused from a landslide in

Tibet in 2000 that ravaged the Northeast and Himachal Pradesh. In this case, while research and policy defined it correctly, there were no institutional mechanism in place that could prevent or at any rate, mitigate it. The lack of an information sharing agreement between the two countries resulted in loss of life, dislocation and extensive damage to property. Both countries have since signed an MOU on the sharing of hydrological data on the Brahmaputra's flows, which will be vital for timely forecasting and management of floods in the Northeast. MOUs signed in 2002 and 2008 provide information on the water level, discharge and rainfall from three stations on the Brahmaputra, namely, Nugesha, Yangcun and Nuxia from 1 June to 15th October every year. [3]

A Missing Water Agenda

What these three illustrative examples underline is the need to shine a light on a range of missing issues that are currently invisible from the mainstream research and policy discourse. For instance, part of the reason why the India-China dialogue on water is a glass half full is that it has ended up being a single-issue debate, fixated on fears over water diversion by China. But has the fixation with diversion diverted policy attention away from other equally critical issues? It has indeed left hiding in plain sight critical issues that do not find a place on the water table. For instance, the fixation with quantity has come at the cost of water quality and drowned out the possibility of a meaningful dialogue on nonpoint sources of water pollution. Even within China, this is becoming a major concern with northern areas including Beijing anxious about the quality of southern waters that will arrive in the north. Added to these are concerns that the fragile ecosystem of the Tibet-Qinghai plateau is showing other signs of stress as it struggles to cope with the furious pace of economic activity that forms part of China's Western Development Strategy. Many mega projects are transforming the face of Tibet. The "pillar" industries of mining and timber processing have fed the rapid industrialisation of Tibet, bringing in its wake the assorted problems of deforestation, soil erosion, landslides, floods, acid rain and pollution especially of the water systems. These are creating ecological imbalances in the form of rising temperatures, retreat of glaciers and droughts caused by indifferent rainfall. Much of these will find their way to parts of the extended region. A particular area of concern for downstream countries could be the environmental degradation facing the Tibetan 'Three Rivers area' comprising the Yarlung Tsangpo, Lhasa River and Nyangchu basins in central Tibet, located in one of the most mineral resource-rich areas of

China. One of the most intensely exploited areas in this region is the Gyama Valley with large polymetallic deposits of copper, molybdenum, gold, silver, lead and zinc. The Gyama valley is situated south of the Lhasa River, one of the five great tributaries of the Yarlung Tsangpo that merges with it 130 kms downstream. The annual runoff in mega deltas such as the Brahmaputra and Indus is projected to decline by 14 per cent and 27 per cent respectively by 2050. This will have significant implications for food security and social stability given the impact on climate-sensitive sectors such as agriculture. These also raise the larger question about the cumulative impact of massive dam-building projects across the entire Himalayan region and the consequences of such intensive interventions in a region that is ecologically fragile and densely populated. The dangers of water accumulation behind dams could also induce devastating artificial earthquakes (Valdiya 1992). There is also disturbing new evidence emerging about dam-induced earthquakes. Recent research by Chinese scientists has revealed that the Zipingpu Dam may have triggered the Sichuan earthquake in 2008 that resulted in 80,000 casualties (Kerr and Stone, 2009). Would India and China care to learn from the Mekong River Commission's institutional history to invest the early decades in data-gathering projects which have contributed to strengthening transparency and, more importantly the notion of water as a regional public good? Another possible template in this regard is the Sino-Russian agreement of 2006, which established an emergency notification system in the wake of the severe pollution of the Songhua River in Russia due to a chemical plant explosion in China's Jilin province.

These only go on to underline the importance of developing a baseline database on water resources within the region. The capacity to predict water flows and water balance in any given transboundary river basin would hinge a great deal on addressing critical gaps in the knowledge base. But recent scientific studies have disturbingly brought out grave errors in current assessments of precipitation. High-altitude precipitation in the upper Indus basin was found to be higher by a factor of 10 than what is currently estimated. Such discrepancies are due to the lack of adequate meteorological stations in upper elevations, with datasets having to rely on a combination of satellite images and available measurements received from stations mostly located in the plains. The cumulative impact of run-of-the-river dams also remains ill defined and little understood. The fact that these will require storage of large volumes of water for power generation has raised serious concerns downstream. Evidence is also emerging that such projects tend to retain substantial silt deposits, resulting in decreased sediment flows downstream. More than a million people in the Ganges-Brahmaputra delta are likely to be affected due to decreased sediment delivery and increased sea level rise by 2050. At the end of the day, while technical issues of measurements, flow patterns and runoffs have their importance, it is more often the more intangible perceptual aspects that create and entrench positions that can produce or retard cooperation at the transboundary level.

Learning to Unlearn

The manner in which these difficult trade-offs are negotiated will decide whether the Kerala floods will be remembered as yet another disaster or if it can constitute a wake-up call for unlearning. The kind of choices that are made will ultimately decide the kind of flood resilience model the state ends up with. This will mean a capacity to not only question but also move beyond problematic models on water such as 'keeping people away from water' or 'keeping water from people', which have had only a mixed record. For instance, 'keeping water from people', the more conventional approach of building structural defences for flood prevention also had a rather mixed record as international experience tells us. While the building of dykes on the Danube, Europe's second largest river has had a somewhat limited measure of success, Vietnam has had a larger scale of dyke failures. In stark contrast to these problematic models is an interesting instance of innovation from Mozambique. Further, Mozambique could make only modest headway in discouraging construction in flood risk areas. In what could point to an alternative paradigm of responding to floods and institutionalising societal learning, is the use of a card game that was used to not just spread awareness among the young but also to instil the idea of living with floods.

Framing disasters principally as a technical problem can thus turn out to be a costly analytical cul-de-sac, shutting out the scope for deliberative engagements with a range of local governance actors with experiential knowledge of responding to floods. By potentially shutting out the prospect of alternative dialogues, flawed frames can end up unwittingly to benefit those who are most vulnerable and are directly at the receiving end of risks. Strengthening non-structural risk reduction techniques can go a long way in reducing loss of life and uncertainties. In particular, community risk assessments offer participatory methods to evaluate the nature and extent of hazards and risks and allows for an effective community engagement in risk governance. Grassroots-based disaster preparedness programmes are already using community-risk assessments as a means to reduce vulnerability and enhance coping capacity to hazard. Citizen biodiversity monitoring systems hold out all-round benefits in terms of accurate knowledge about the state of biodiversity, tracking variations based on observations at multiple sites both critical for improved decision making and more effective implementation (Couvet 2008). Importantly, informal social institutions need not necessarily replace formal institutions; rather they can play a supplementary role to bolster service provision. They can in fact perform a critical buffering function of oversight and supervision ceded by the state. They also hold the potential to reduce information asymmetries and enhance governance. Community networks can thus work with, and not necessarily be at cross-purposes with the state on public goods provision.

By and large, there tends to be disproportionate attention given to response strategies during a disaster while the 'pre' and the 'post' phases remain largely neglected. An effective disaster response strategy has to also rectify the intrinsically skewed nature of privileging technical solutions. A community-based flood risk management could go a long way in redressing this dominance of structural solutions (Blaikie et al 1994, Lebel and Sinh 2009). A case in point is the successful community managed flood management model of Bangladesh that effectively reduced vulnerability during the 2004 floods. Bringing in perspectives from outside the formal government structure and enabling participation from below can also help voice silences. For instance, volunteer and community groups can play in a range of disaster prevention from early warning, disaster response and post-disaster response. Informal volunteerism is likely to provide a much-need 'surge capacity' needed for disaster response and recovery. The role played by fishermen in Kerala flood rescue efforts is a case in point of the effectiveness of informal volunteerism at work. Similarly, BlazeAid, a community organization set up in the aftermath of the Black Saturday bushfires in Australia in 2009 provided crucial post-disaster assistance in the form of rebuilding rural infrastructure damaged by floods and other natural disasters. 70,000 volunteers took part in the massive post-flood clean-up drives in Kerala, which included cleaning debris from around one lakh buildings (First post 2019). Bangladesh has a 50,000-strong active volunteer force under its Cyclone Preparedness Programme that covers 13 coastal districts. A related new trend emerging is the rise of digital volunteerism, which is also helping to plug crucial gaps in flood risk response. An illustrative example is the live crisis map that graduate students of Tufts University produced in the immediate aftermath of the 2010 Haiti earthquake that helped identify the scale and extent of damage. Communities play an active role in the planning and implementation of water, irrigation and sanitation projects in Brazil and Taiwan, representing a successful instance of what John Ackerman refers to as 'co-governance' (Ackerman 2003).

By its very nature, approaches to enhance flood resilience will call for balancing the technical with the social perspectives. The social dimension is particularly critical since steps taken towards mitigation and prevention may have quite unintended pre-flood harm impact in the form of loss of income or devaluation of property (van Doorn-Hoekveld et al 2016). Since disaster response is crucially about societal regeneration, re-engaging with the social has to be the first step towards people-centric adaptations. Such a reorientation will also recognise society as a shaper of knowledge and not as a mere passive receiver. It is also vital to keep in mind the distributional effects of flood risk management. This can introduce vulnerability as a key concept and the 'capacity to be wounded' by looking at how different social groups are affected by policy interventions. By thus localising impact studies, it can allow for an inherently place-centric notion of socio-economic stress and impact. It is only when a diverse set of actors engage in extensive interactions and learn from such interactions that these processes can produce socially inclusive development outcomes.

One cannot overstate the importance of early action in flood response measures such as ensuring safe water supply, food, shelter and medical care. In Mozambique in 2000, severe floods were followed by rioting due to a widespread public perception of an abdication of responsibility on the part of the government. Logistical inefficiencies and coordination challenges in relief distribution further compounded reconstruction efforts in Haiti. Lack of sanitation, the mixing of sewage with floodwaters, risk of water-borne diseases such as cholera presented further grave challenges in the Haiti earthquake response. It is to Kerala's credit that there were no reports of outbreak of communicable diseases from the flood-ravaged areas. A great deal of the credit for this owes to the effectiveness of the state's healthcare system with decentralised monitoring at district levels playing an important part in averting major outbreaks.

But while effective disaster response is in and of itself critical, long-term measures that can play a vital role in preventing floods are no less important. These include measures to ensure erosion control, soil stabilisation, and drainage systems for irrigation among others (IEG 2010).

Flood risk governance experience in Europe shows the increasing importance governments are giving to nature-based solutions to reduce run-off. These include protecting wetlands, widening natural flood plains and restoring reefs. Such adaptation measures have reportedly reduced the damage from Hurricane Sandy in the US by an estimated $625 million (Akpan 2017). Similarly, China's Sponge Cities programme brings the focus on imaginatively integrating green infrastructure to urban design. Another innovative disaster response is the manner in which the Philippine Red Cross used social media tools such as Twitter not just to locate flood sites but also to understand public perception and garner feedback. Flood risk maps can also be a useful policy tool to assess the magnitude of risk that people are exposed to by looking at the dimensions of health, livelihood and environment.[4]

Beyond Theoretical Navel-Gazing

The intellectual discourse on water too has been problematic because of its inability to project alternative scenarios as well as raise critical new questions. A straitened debate on water and an often-uncritical intellectual acceptance of this has unfortunately diminished the space for unlearning. Within these structural parameters, the role of people has thus been reduced to being passive recipients of knowledge flows. The mainstream thinking on water has tended to remain in a state of denial despite mounting evidence to the contrary. Social science research on water issues in South Asia has typically tended to focus on macro-level water issues such as droughts, floods, rivers, dams, irrigation etc. and have limited end-user involvement and end-user needs or problems. Issues of regional planning regarding rivers that are transboundary in nature often do not find a place in the research agenda and analysis. Studies on rivers avoid taking a comprehensive view of the river system as a whole. The river is often studied in isolation with little attempt to relate it to the ecological system of which it forms an integral part.

Unless some of these dominant interpretations are called into question, extant categories and conventional imageries will only be reified over time without respite. Synergies between science and traditional knowledge can for instance bring actionable ideas that can create bottom-up, inclusive approaches to questions of resource governance. Similarly, building bridges between research and policy will also be critical to create institutional entry points for collective learning. Such an approach can, as John Dryzek points out, offer valuable lessons in social learning "for it does not rule out a variety of experiments in what sustainability can mean in different contexts, including the global context" (Dryzek 1997). This introduces a spatial dimension and a multiplicity of actors that go missing in meta-narratives. Such a complex systemic understanding recognises the validity of place-centric, contextual knowledge. Multiple pathways for learning are significant since "tacit knowledge is difficult to 'transport'... [and] tends to accumulate in specific places" (Rohracher 2006). It will above all create entry points for experiential learning and place context at the centre of the learning process. Decontextualised knowledge cannot convey the lived experiences and its capacity to community ownership gets severely limited (Duffy and Jonassen 1992). This will allow for multiple scales, levels and actors and their iterative interactions. Rethinking scale as a category of practice, would as Brubaker and Cooper argue, define it as 'categories of everyday experience, developed and deployed by ordinary social actors' as against 'experience-distant categories used by social scientists'.

This will in turn mean not just problematising ways in which knowledge is produced and utilised in society but is in a fundamental sense also about what counts for knowledge. Recognising that 'there may be several valuable ways of knowing' has to form the basis of any interdisciplinary enquiry. This will also mean consciously eschewing "bipolar models of knowledge use" which as James Boggs notes, defines knowledge essentially as a transaction between two parties- "knowledge producers (social scientists) and knowledge users (eg., policy makers or decision makers)" (Boggs 1992 : 29). Such a dyadic framing of knowledge fails to recognise people as important repositories of knowledge, denying them agency as social actors of their own merit. The mainstream notion is that knowledge and research have to in the ultimate analysis, be relevant and useful to policy makers as their end user has dictated choices and predetermined terms of inclusion and exclusion. Thus, the need for multi-actor frameworks involving local communities, researchers, policy makers and development workers is all too evident. But what is arguably even more important is to get syncretic water conversations going among these diverse stakeholders. The notional distinction between research and the researched for instance has meant that research findings and local knowledge often talk past each other. The perception that research has to, in the ultimate analysis, be relevant and useful to policy makers has further constricted knowledge frames. These also call for governance arrangements to not only jump scales but also make the conceptual leap towards co-governance for more effective flood risk communication and coordination.

The 2018 Kerala floods lend an urgency to begin a broader conversation on India's resource choices, its drivers as well as determinants and its likely consequences for the life chances of local communities who tend to be at the receiving end of flood fury. This is essential since flood risk governance is fundamentally a crosscutting concept, which cannot be seen in isolation from critical issues of poverty reduction, land-use management, water supply, solid waste management and disaster prevention. Such a disciplinary turn could create valuable entry points for experiential learning (and unlearning) as well as for a whole set of missing issues that are currently invisible to the mainstream policy and research gaze.

Notes

1. Rising temperatures are widely expected to bring alterations in the hydrological cycle and patterns of rainfall, affecting precipitation as well as runoff. A 3-5 metre rise in average sea level holds implications for coastal cities such as Dhaka and Kolkata with large population bases (Byravan and Rajan, 2008).
2. They were so called owing to the suitcases made of cheap carpet material that they carried. For further details, see Heather Cox Richardson (2001): Death of Reconstruction: Race, Labor, and Politics in the Post-Civil War North, 1865-1901. Harvard University Press, Cambridge.
3. Each MOU is for a duration of five years and has to be subsequently renegotiated for continued release of data. The two countries have also set up the Joint Expert Level Mechanism, a decision taken during Hu Jintao's visit to India in 2006.
4. The Implementation Plan on provision of hydrological information signed by India and China in 2014 also includes the provision for site visits. According to Article 12 of the Implementation Plan, "In order to ensure normal provision of hydrological information, if necessary, after mutual consultation through diplomatic channel, the parties may dispatch hydrological experts to each other's country to conduct study tour according to the principle of reciprocity" (Ministry of External Affairs 2014).
5. The European Union Floods Directive makes it mandatory for all member countries to produce flood maps to create a region-wide 'framework for the assessment and management of flood risks' (EU Floods Directive 2007).

References

1. Ackerman, John. (2003): "Co-Governance for Accountability: Beyond "Exit" and "Voice", *World Development*, 32 (3): 447-463.
2. Ahmed, A U, K Zahurul K Prasad, S N Pudel and S K Sharma (2004): "Synthesis of manuals on community floods management in BD, India and Nepal", *Asia Pacific Journal of Environment and Development 11.*

3. Akpan, Nsikan, (2017): "Wetlands stopped $625 million in property damage during Hurricane Sandy", *Public Broadcasting Service, accessed* at https://www.pbs.org/newshour/science/wetlands-stopped-650-million-property-damage-hurricane-sandy-can-help-houston
4. Blaikie P T, Cannon I Davis and B Wisner (1994): *At risk: Natural hazards, people's vulnerability and disaster*, Routledge, London
5. Boggs, James (1992): "Implicit Models of Social Knowledge Use", *Science Communication*, vol. 14, no.1.
6. Brubaker R., and Cooper, F., (2000): "Beyond identity", *Theory and Society* 29.
7. Craig, Gary (2007), "Community Capacity-Building: Something Old, Something New...", *Critical Social Policy*, vol. 23, no.3.
8. Dev, Rajesh (2007): "Negotiating Diversities through Institutional Strategies", *Eastern Quarterly,* Volume 4, Issue I, July - September.
9. Dowsley, Martha (2008): "Developing Multi-level Institutions from Top-down Ancestors", *International Journal of the Commons*, Vol 2, no 1, January.
10. Dryzek, John, (1997): *The Politics of the Earth: Environmental Discourses*, Oxford University Press.
11. Dryzek, John S. (2002): "A Post-Positivist Policy-Analytic Travelogue", *The Good Society,* Vol. 11, No.1.
12. Duffy, T.M. & Jonassen, D. eds. (1992): *Constructivism and the technology of instruction: A conversation*. Hillsdale NJ: Lawrence Erlbaum Associates.
13. European Union Floods Directive (2007): *accessed in http://ec.europa.eu/environment/water/flood_risk/implem.htm*
14. Firstpost. (2019): https://www.firstpost.com/india/kerala-after-the-flood-70000-people-volunteer-for-clean-up-drive-in-kuttanad-aim-to-clear-debris-from-1-lakh-buildings-5063441.html
15. Ghani, M.U.: "Participatory Strategy for Flood Mitigation in East and Northeast India: Case study of the gangesbrahmaputra-meghna basin, Farakka Dam Project", *Ministry of Water Resources, Govt of India* at http://209.85.175.104/search?q=cache:DzCRjTxvngoJ: www.unescap.org/esd/water/disaster/2
16. Huang X, et al, (2010): "Environmental impact of mining activities on the surface water quality in Tibet: Gyama valley", S*cience of the Total Environment.*
17. IEG (2010): *Responding to floods in West Africa: lessons from evaluation*. Washington, DC:IEG-World Bank.
18. Innis, Harold (1950): *Empire and Communications*, Oxford: Clarendon Press
19. Intergovernmental Panel on Climate Change, (2007): *Contribution of Working Group I to the Fourth Assessment Report of the* IPCCaccessed athttps://www.ipcc.ch/publications_and_data/ar4/wg1/en/contents.html
20. Klein, Naomi. (2007): *The Shock Doctrine: The Rise of Disaster Capitalism*, Toronto: A. A. Knopf, Canada.
21. Kurian, Nimmi (2018): "Uncharted Flows: Navigating the Downstream Debate on China's Water Policy", in in K. J. Joy et al, *Water Conflicts in Northeast India*, Routledge.

22. Kurian, Nimmi. (2017): “Addressing the Drought of Ideas on the Brahmaputra”, *China-India Brief*, Lee Kuan Yew School of Public Policy, National University of Singapore, 24 October.

23. Lebel L and B T Sinh (2009): “Risk reduction or redistribution? F M in the Mekong region”, *Asian Journal of Environment and Disaster Management,* 1, 23-39.

24. Lundy, Mark, Marcia Veronica Gottret and Jacqueline Ashby (2005): “Learning Alliances: An Approach for Building Multi-Stakeholder Innovation Systems”, *ILAC Brief*, August.

25. Mishra, Dinesh Kumar (2011): “The people vs the environment”, *Himal*, March.

26. Newell, Peter and Joanna Wheeler (eds.) (2006): *Rights, Resources and the Politics of Accountability*, Zed Books, London.

27. Newell, Peter and Joanna Wheeler eds. (2007): *Rights, Resources and the Politics of Accountability*, Zed Books, London.

28. Rammel, Christian et al., (2007): “Managing complex adaptive systems-A co-evolutionary perspective on natural resource management”, *Ecological Economics*, 63: 9-21.

29. Richardson, Heather Cox (2001): *Death of Reconstruction: Race, Labor, and Politics in the Post-Civil War North*, 1865-1901. Harvard University Press, Cambridge.

30. Rohracher, Harald. (2006): “Governing Sociotechnical change in regional innovation systems” in Tchalakov I, Mali M, Rohracher H (eds), *Governing Sociotechnical change in South-Eastern Europe*: *Contributions from a science and technology studies perspective, East-West Publishers, Sofia.*

31. *Times of India*, (2017): “Causes behind Brahmaputra turning black could be natural: Union minister”, 4 December.

23

Understanding Kerala (2018) and Himalayan Floods (2013-14)

Shekhar Pathak

Floods have been part of natural systems of the planet earth from earliest times, like earthquakes, landslides, avalanches, hurricanes and tsunamis. After the beginning of agriculture and urbanization the direct human contributions started associating with it. Floods can change the geo-morphology of a region, shift the soil and change the landscape drastically. Floods depend on so many factors- the natural hydrological systems, the rains (monsoons in Indian context and double monsoons in Kerala context), the state of ground water, the state of the catchments, the slopes, the vegetation, the seismicity of the region etc. The floods also depend on the volume of the rains and their erratic nature, cloud bursts, dams and barrages like man-made obstructions.

Humans learnt from the floods since the very beginning of human history on the earth. They always hesitated in settling in the flood plains or river courses. They have been keen observers of the river behaviour, change in the water volume or colour. In the Himalayan context the communities have understanding of the behaviour of glaciers or events like glacial lake bursts. But humans are also great encroachers and they sometimes disobey the rules made by themselves in the interest of all. At that point we need appropriate implementing agencies, which can follow the accepted appropriate traditional ways and land, water related laws of the country to reduce the consequences of the calamity.

The smaller floods have been creating plain areas by bringing fertile soil from the mountains. Many a time the floods have created 'land benches' along the rivers, which later found suitable for human settlements and agriculture. As Indian sub-continent has many geological, climatic, agro-ecological regions, so the rivers and floods have different characteristics in different regions. This paper examines the Kerala flood (2018) and compare it with the Himalayan floods (2013-2014).

2. Looking at Kerala

Kerala is situated in the South-Western corner of Indian subcontinent, bordering two Indian states of Tamil Nadu and Karnataka, between the mountains of Western Ghats[1] and the Arabian Sea. The province of Kerala, created after the States Reorganization Act of 1956, is spread over 38863 square km. with 33387677 populations in 2011. Divided into 14 districts Kerala is 12th largest state economy in India with a composite culture.

Between Western Ghats (Sahyadri mountain range) and Malabar Coast (Lakshadweep Sea) Kerala has a unique ecology with high mountains, gorges and deep cut valleys in eastern region; hills of less height and lowlands in the central region and coastal plains in the western region. The western ghats has an average height of 1,500 meters and the highest peak is 2,500 meters. The mountain range and coast line run parallel. Due to this Kerala has humid tropical climatic conditions. The ecology created by this climate is the cause of Kerala's rich biodiversity. Kerala has a coastal line of 590 km and maximum width of the state is 121 km and minimum is only 11 km.

Kerala has a total 44 small rivers and out of them 41 are west flowing and only 3 are east flowing. Periyar (244 km) is the longest river and other important rivers are Bharathapuzha (209 km), Pamba (176), Chaliyar (169), Kadalundipuzha (130), Chalakudipuzha (130), Valapattanam (129) and Achankovil (128 km). The average length of the rivers is only 64 km. These rivers make eight percent of India's internal waterways. Many of the rivers are small, spring fed and many are only monsoon fed. The deltas are invisible phenomenon here. The changes in land use, sand mining and quarrying have been accelerating the land-mud slides, floods and droughts for some decades.

3. The Double Monsoon Ecology of Western Ghats

Kerala is the southern part of western ghats. The 'Western Ghats', mountain system is spread across many states, refers to the practically unbroken hill chain (with the exception of the Palakkad/Palghat Gap) running roughly in a north-south direction, for about 1500s km parallel to the Arabian sea coast, from the river Tapi (about 210 16' N) down to just short of Kanyakumari (about 8019' N). In some accounts the term Western Ghats or Sahyadris is restricted only to the western escarpment of the Peninsular Plateau from the Tapi southwards to the region of Kodagu, (about 12 degrees N) while the higher mountain ranges further south, including the Nilgiris, the Annamalais, the Cardamom hills and the Agasthyamalai range, being referred to as a distinct geological entity named as the Southern Block. We use the term 'Western Ghats' in the broader sense to include the entire tract of hills from the Tapi to Kanyakumari, with an area of approximately 129037 sq km, it stretches to a width of 210 km in Tamil Nadu and narrows to as small as 48 km in Maharashtra (leaving the Palghat gap)- (Gadgil Report, pp: 5-7). The Ghats host India's richest wilderness in 13 national parks and several sanctuaries and is recognized by UNESCO as one of the world's eight most important biodiversity hotspots. The ecosystems of the Western Ghats comprises parts of six states - Gujarat, Maharashtra, Goa, Karnataka, Kerala and Tamil Nadu. The total area of Western Ghats is 129037 sq km, total length is 1490 km, minimum width is 48 km and maximum is 210 km.

The double monsoon climate ecology had been functional for long in Kerala. The total rainy days annually are 120-140. Average rainfall is 2923 mm annually and 65 % rains happen between June to August and rest between September and December. If we look at the history of disasters in Kerala we don't have a long list, like we have in Hindukush-Himalaya. The great flood of 1341 AD is often mentioned while talking about the calamities in Kerala. It is told that this mid 14th century flood drastically changed the terrain, many river courses and the overall landscape of the region. But we don't have details of that calamity. Was it related to any earthquake and tsunami? Was it a combined one? We don't have clear answer.

We know more about the second big calamity that happened after 583 years in 1924. It was documented that in 1924 a combination of heavy rains and breaching of the dams triggered the devastating floods causing more than 1000 deaths. The 2018 floods are called 'the repetition of this tragedy' as it is also the result of the heavy rains and bad dam management. The data from the Indian Meteorological Department (IMD) shows that Kerala received 2,087.67 mm of rain from June 1 to August 15 in 2018, which is 30 percent less from 3,368 mm rain in 1924. The death toll in 2018 was close to 500, the loss of property 20000 crore and nearly one and a half million people were affected.

What drastically changed in last 94 years after the 1924 floods? One answer is change in demography and increased pressure on the resources of the land. The second is sudden change in raining pattern and the increasing number of cloud bursts. The third is failure in managing the newly built dams. The fragility of the Western Ghats is common in all three points. Tectonic and seismic activities records should also be looked at. It is to be noted that Kerala state has the lowest population growth rate in India. It is only 3.44% a year. It has the highest *Human Development Index* (HDI), 0.712 in 2015; the highest literacy rate, 93.91% (2011 census); the highest life expectancy of 77 years; and the highest sex ratio (1,084 women per 1,000 men).

The twin issues of the pressure on natural resources and the mismanagement of the dams are to be examined in a more critical way. In the last hundred years the development of roads, dams; rise in urbanization; expansion of coffee, tea and rubber plantation; increase in resorts and high rise buildings in the mountainous and wild areas and rapid growth in sand mining and quarrying have destroyed the biodiversity, green cover and over all wilderness of the Western Ghats. A study in the southern region, comprising the states of Karnataka, Kerala and Tamil Nadu, showed that between 1920 and 1990 about 40% of the original vegetation cover was lost or converted to another form of land use (Menon and Bawa 1997, Gadgil: 9). The State of the Forest Report, 2017 also clearly tells us that the reason for the increase in the forest cover is commercial plantation, not the preservation of natural forests (India State of the Forest Report 2017, FSI, Dehradun: 223). Plantations increase the greenery but cannot make a natural forest.

Normally it is thought that every dam can moderate floods in the downstream areas, as long as it has space to store water. Every action which helps in storing, holding/delaying the flow of rainwater from the catchment to the river moderates its flow and the floods. Today the catchments of rivers are fast losing this capacity, due to deforestation, encroachment into wetlands, local water bodies and fast changing land use. This process also reduces soil's capacity to hold water and delay its flow.

Kerala's 44 rivers with short lengths and high slopes have over 60 large dams. The potential capacity of dams to help moderate floods can be realized only when they are operated with this objective in mind. When dams are not operated with such an objective, and are, instead, filled up as soon there is water available, there is no space left to store more water. The only alternative then is to release all the inflow into the downstream river. Due to this, in downstream areas, which are already facing floods due to local rains or other reasons, the dams end up increasing the magnitude of the flood disaster. Himanshu Thakkar has mentioned numerous instances of this kind, including the floods in Uttarakhand (June 2013), Tehri (September 2010), Hirakud (2009, 2011, 2014), Damodar dams (multiple years), Krishna basin dams (2006, October 2009), Ukai (August 2006), Chennai floods (December 2015), Bansagar dam (August 2016), Kurichu dam in Bhutan (2004, 2016, others), and Ranganadi (2017) and Doyang (2018), among other dams, where flawed operation of the dams created or worsened flood disasters in the downstream areas (SANDRP and EPW, September 22, 2018).

There was definitely a difference between the pre-dam and post-dam floods. At the time of 1341 flood there were no dams in Kerala Rivers and in 1924 there were some dams built. Here we should know the difference between the pre-monsoon and post-monsoon floods. Kerala experiences South-Western summer monsoon and North-Eastern winter monsoon. There is no dispute that in Kerala, most of the dams were almost full by the end of July 2018. The Central Water Commission (CWC) report on the Kerala floods, however, stated that the dams in Kerala neither added to the flood nor helped in reducing the flood, as most of the dams were already at FRL (Full Reservoir Level) or very close to FRL on 14 August 2018, due to more than normal rainfall in the months of June to July 2018 (CWC 2018: 34).

But Himanshu Thakkar questions this analysis by terming such statements :

> an effort to absolve the dams from any blame, assume and imply that the floods in the river downstream, before and after the dam, are same. This is a misleading and incorrect assumption. The flood in the downstream area, from water released by the dam, is very different from the flood in an undammed river for a number of reasons. First, floods in rivers before the dam can, most of the time, be seen gradually rising, which allows people to prepare and can save lives as well as valuables. Floods from water released by the dams come much more suddenly, leaving little time to respond. Second, the potential damage from water suddenly released from a dam is much greater than that of a gradually rising flood in the river. Third, the river downstream, its flood plains, and even the riverbed - which do not experience regular

floods - may have changed. The dam induces a false sense of security from floods, often propagated by dam proponents, which leads to encroachment on the river bed/flood plain. The carrying capacity of the downstream river may have also changed. The river immediately downstream of the Idukki dam saw no flows for 26 years; the last time the gates were opened was in 1992.

The flood-carrying capacity of the Tapi river, downstream of Ukai dam, has reduced from about 8.5 lakh cubic feet per second (ft^3/sec) earlier, to about 4.0 lakh ft^3/sec, and to possibly less than 2 lakh ft^3/sec at present (CWC 2005: 46; Patel and Gundaliya 2014). Finally, the water released from the dam is either relatively silt-free or carrying too much silt if released from the bottom sluices. In either case, it is different than normal floodwater, and has a different damage potential. Therefore, dam operators cannot be exonerated on the claim that they have not added to the incoming floods.

The real question is: Have the dam operators done everything possible to moderate the floods in the downstream areas? The CWC report does not even ask that question. So, what could the dam operators have done to moderate the floods in Kerala during the 2018 monsoon?"

Himanshu Thakkar further writes about violation of the 'Rule Curves':

> Every dam is supposed to have a dam-specific rule curve that regulates, among other things, how the dam is supposed to be filled during the monsoon to optimize flood moderation for the downstream areas, while ensuring that it is filled up only towards the end of the monsoon. Almost all the dams in Kerala were full by the end of July. This was in complete violation of the rule curve, since by the end of July the south-west monsoon is just halfway through its course, and large parts of Kerala also receive the north-east monsoon, which follows the south-west monsoon. Thus, to fill up the dams by end of July was an invitation for disaster.

The CWC report is clearly designed to absolve the dam operators from any blame, so it does not mention whether the Kerala dam operators followed or violated the rule curves. However, the very first recommendation of the report is: It is essential to review the rule curves of all the reservoirs in Kerala. The rule curves need to be formulated for both conservation as well as operations during the flood, particularly for the reservoirs having the live storage capacity of more than 200 MCM (million cubic meters) in order to create some dynamic flood cushion for moderating the floods of lower return periods particularly in the early period of monsoon (CWC 2018). The Kerala dams clearly failed in flood moderation during the August 2018 floods, as they were already full when the floods occurred" (EPW: September 22, 2018).

Around 5254 dams in India are an integral part of flood management, apart from storing water for irrigation and generating power. A report of the Comptroller and Auditor General (CAG) submitted (2017) in Parliament stated that there was emergency action plan available for only seven percent of these dams. For the 61 dams in Kerala, there is none. The oldest dam of Kerala Mullaperiyar Dam was built in 1895 on the river Mullayar and its tributary, Periyar to provide water to deficit regions of Tamil Nadu - especially arid shadow region of Madurai which was in live need of water. That is why the Periyar River which flows westward of Kerala was diverted towards the Bay of Bengal to provide water to the people of Madras presidency. This dam is owned and operated by Tamil Nadu Public Works Department. In 1979 the Kerala government sought its rebuilding due to its being in earthquake prone area. In 1980 Tamil Nadu government demanded to raise its height from 136 feet to 150, which was opposed by Kerala. In 2006 the Supreme Court directed to raise the height to 142 feet and Kerala brought a law to negate it. After that the Supreme Court appointed a panel to examine the safety and the possibility of a new dam. The other major dams are Idukki, Idamalayar, Kakki etc. More than half of Kerala's dams (around 57%) are hydro-electric projects operated by the Kerala State Electricity Board and the rest are operated by the irrigation department of the state. For both entities, the amount of water to store is motivated by demand for electricity and irrigation, rather than flood control measures.

The 2017 CAG report presented an insight on what is wrong with India's flood management plan. Of the 219 proposed new telemetry stations, used to forecast floods, only a quarter, were set-up till August 2016. Of the 375 existing stations, almost 60% were non-functional after installation. The flood forecasting data was used in the formulation of flood forecast only after comparing the telemetry data with manually observed data; and in the case of mismatch between the two sets of data, manual data was adopted. CWC could not use and depend on telemetry data even after investing in modernization for nearly 20 years. The report also said that there is an emergency action plan for only 7% of the dams in India. For the 61 in Kerala, there is none as told above. In eight out of

17 flood-prone states, the integrated flood management plans for entire river/basin have not been taken up, the report said.

A recent study also warns us. Earthquakes of similar magnitude of 2004 Sumatra earthquake could occur in an area beneath the Arabian Sea at the Makran subduction zone, according to recent research. The study suggests that the risk from undersea earthquakes and associated tsunami in this area of the Western Indian Ocean, which could threaten the coastlines of Pakistan, Iran, Oman, and India and further, has been previously underestimated. The results highlight the need for further investigation of pre-historic earthquakes and should be fed into hazard assessment and planning for the region.[2]

Subduction zones are areas where two of the Earth's tectonic plates collide and one is pushed beneath the other. When an earthquake occurs in these areas the seabed moves horizontally and vertically as the pressure is released, displacing large volumes of water that can result in a tsunami. The Makran subduction zone has shown little earthquake activity since a magnitude 8.1 earthquake in 1945 and magnitude 7.3 in 1947. Because of its relatively low seismicity and limited recorded historic earthquakes it has often been considered incapable of generating major earthquakes. Plate boundary faults at subduction zones are expected to be prone to rupture generating earthquakes at temperatures of between 150 and 450 C. The scientists used this relationship to map out the area of the potential fault rupture zone beneath the Makran by calculating the temperatures where the plates meet. Larger fault rupture zones result in larger magnitude earthquakes.

The other and hidden aspect of future calamity calculation is related with climatic changes happening around us without noticed by us seriously. Climatic changes are happening all over the planet earth. At some places these are little visible, at other places these may be seen in larger form so more visible to us.

4. What Western Ghat Ecology Experts Panel Report Suggests

Now let us look at the Western Ghat Ecology Experts Panel Report (2011). The WGEEP Report chaired by Madhav Gadgil has suggested so many things to be done in Western Ghats and Kasturirangan's 'counter' Committee (2017) has not opposed all of them (the government did the same thing by forming another committee in the case of Ravi Chopra Committee initiated by Supreme Court on the Dams in Uttarakhand). Even if we come to the golden mean of both the reports (Gadgil and

Kasturirangan) an implementable point can be identified and strategy can be evolved.

But the politicians with some vested interest, as everywhere turned it into a debate of development v/s environment. In essence WGEEP advocated a layered, nuanced, participatory approach, so that boundaries will not be discontinuities and therefore will not be of undue significance. The report talks of the boundaries of the Western Ghats, but pleads that the pattern of adaptive co-management that we propose may also be applied to regions beyond these boundaries (page 15-16).

The panel has given the idea of Ecologically Sensitive Zones-ESZ (page 23) and for overall management it has suggested to establish Western Ghats Development Authority (WGDA) and Western Ghats Ecology Authority (WGEA) covering the whole region spread in five states. The report has suggested three types of ESZs and ESA (Ecologically Sensitive Areas). But the idea is old. Western Ghats as an administrative entity was first visualized in the context of Regional Planning exercises, beginning with a report prepared by the Town and Country Planning Organization, Delhi in the 1960s. This report delineated the Western Ghats at Taluk level, which became the basis of the Planning Commission of India's Western Ghats Development Programme (WGDP) initiated in 1974-75 across 132 talukas. This serves as the basis of disbursement of central government assistance. The WGDP is currently being implemented in 171 talukas of Western Ghats viz. Maharashtra (63 talukas), Karnataka (40), Kerala (32), Tamil Nadu (33) and Goa (3 talukas). Now some of the original talukas have been sub-divided.

It is interesting that the Kerala Bio Diversity Board (KBDB) has also given its ideas tuning with some of the panel's recommendations :

> Water resources management in the Western Ghats region is linked to improving the flows in the rivers and the health of the catchments. Shorter perennial monsoon fed west flowing rivers like Periyar, and the Bharathapuzha travel through steeper and more undulating topography before emptying into the Arabian Sea. As for Kerala, the groundwater potential is low when compared to other states and shallow dug wells are the most common source of freshwater. However, over the years the groundwater table is lowering at an alarming rate indicative of poor recharging capacity. On the other hand, water needs are growing in the Western Ghats States. In Peechi and Malampuzha reservoirs irrigation water is being diverted for drinking and for

> the industrial needs of cities in Thrissur and Palakkad respectively. As for the west-flowing rivers, saline ingress is advancing even into the midlands due to reduced downstream flows. Crop losses and saline water intrusion into drinking water has been reported in Kerala during severe summer owing to salinity intrusion. The west-flowing shorter rivers (Sharavathi, Periyar) have been dammed at several places. River flows has been completely diverted at Mullaperiyar and Parambikulam dams involving Kerala and Tamil Nadu. West flowing rivers have been virtually made into east-flowing rivers by violating all natural laws. In the case of inter-basin water diversions, absolutely no natural flows or even 'minimum flows', leave alone environmental flows, are left below the dams. In Mullaperiyar dam the main tributary of Periyar has been completely diverted to the Vaigai basin in the east. Idukki dam does not even have a spillway for allowing monsoon spills into the river. Idukki dam is a case wherein the entire catchment was encroached along with dam construction. A river basin-level planning and decentralized management of water resources in the Western Ghats is necessary (Understanding Report of the Western Ghats Ecology Expert Panel, Kerala Perspective, Kerala Bio-Diversity Board: 29).

As per the WGEEP in Kerala Mandakol, Panathadi, Paithal Mala, Brahmagiri-Thirunelli, Wayanad, Banasura-Kuttiyadi, Nilumbur-Mepadi, Silent Valley - New Amarambalam, Siruvani, Nelliampathy, Peechi-Vazhani, Athirappilly-Vazhachal, Pooyamkutty Munnar, Cardamom Hills, Periyar, Kulathupuzha, Agasthya Mala, are identified as ESAs surrounding Protected Areas (WGEEP:20).

The democratic institutions are well entrenched in the state and Kerala leads the country in capacity building and empowering of Panchayat Raj Institutions. So Panel's suggestion not to stick in a system that forcibly divorces conservation from development is fine. The Key features include: focus on learning-by-doing, synthesizing different knowledge systems, collaboration and power-sharing among community, regional and national levels and management flexibility. These features can promote an evolving, place-specific governance approach in which strategies are sensitive to feedback (both social and ecological) and oriented towards system resilience and sustainability. Such strategies include dialogue among interested groups and actors (local-national), the development of complex, redundant and layered institutions, and a combination of institutional types, designs and strategies that facilitate experimentation and learning through change. Other important themes in adaptive co-management include improving evaluation of process and outcomes, additional emphasis on power, the role of social capital, and meaningful interactions and trust building as the basis for governance in social-ecological systems (WGEEP, 2011 page 15-16).

Yet we are today stuck in a system that forcibly divorces conservation from development. It ends up creating a dichotomy so that our policies at once promote reckless development in certain areas and thoughtless conservation in other areas. In the process we constitute islands of biodiversity (and social exclusion) - the so-called Protected Areas (PAs) - in an ocean of ecological devastation outside of these PAs.

For the Kerala state within Western Ghats the total number of proposed ESZ1 is 15, ESZ2 is 5 and ESZ3 is 8 (WGEEP: 25). Together with it are the Eco Sensitive Areas. Out of Western Ghats' total landscape (129037 sq.km.), 77000 sq.km. (including existing sanctuaries) will come under SEZ1, which will be 60% of the landscape. 20000 sq. km. (15%) will go to ESZ2 and 32000 sq. km. (25%) will go to ESZ3 to allow all developmental activities with precaution. The report did not find favour with industry and concerned states. Instead of taking decision the MoEF appointed another panel, High Level Working Group (HLWG) led by K. Kasturirangan to examine the Gadgil report, consult the stakeholders and suggest the ways to implement it. The report of this panel came out in April 2013 with the suggestions of its own on ESAs. HLWG followed the criteria of the Western Ghats Development Programme of the Planning Commission and identified 188 talukas as its Western Ghats landscape, which worked out to 1,64,280 sq km. The report marked 37 percent (60000 sq km) of this stretch as ESA where hazardous industries, thermal plants or mines would not be allowed. There has been lot of debate on ESAs suggested by Gadgil and Kasturirangan and many environmentalists are insisting on acceptance of ESAs suggested by Gadgil.

The WGEEP has given so many suggestions regarding 'no go' areas, power plant clearance, rail and road development, waste disposal etc also. It looks at the issues of Western Ghats in a complete way.

5. The Case of Uttarakhand Floods

The two Himalayan floods should be examined in a larger canvas. In June 2013, heavy rains and *cloudburst* caused devastating floods and landslides in *Uttarakhand*. Though the IMD informed the people through its bulletins that the monsoon would reach in the last week of June these were not seriously taken by pilgrims, administration and government at large, causing thousands of people to be caught unaware, resulting in huge loss of life and property.

Two days of continuous rains resulted in the collapse of the Chorabari lake and landslides and the floods caused havoc in the whole of Uttarakhand, parts of Himachal (Kinnaur-specially Baspa valley), Western Nepal and Western Tibet. The natural silt which comes with the water of the rivers from glaciers, debris created by faulty road building (dynamiting and debris throwing into the rivers), project constructions and the muck deposited by hydro-companies near river banks accelerated the flood. The collapse of many hydro projects in different valleys further increased the devastation. The debris blocked up the rivers, causing major overflow. Some parts of Far West Nepal, *Haryana, Delhi* and *Uttar Pradesh* also experienced heavy rains. Around 6000 people, as per official records, were presumed dead. The people killed during the devastation include locals, Nepali labourers and mostly pilgrims from all over the country. The actual number of people who lost their lives may be more than ten thousand as the Chairman of Uttarakhand Vidhan Sabha reported.

The destruction of roads, bridges, buildings, hydro-projects, canals, power houses, vehicles was so much that some experts say that the overall loss was higher than what Uttarakhand lost in the 1991 earthquake. More than 300000 pilgrims were struck in Chardham (Badri, Kedar, Gangotri, Yamunotri and Henkunt Sahib) region. Many died due to rains and cold, many due to landslides and floods. Many lost lives, while running towards high places in Kedar valley. Search for bodies who died during the extreme natural conditions continued for several months and even as late as September, 2013, about 556 bodies were found, out of which 166 bodies were found in highly decomposed state. Later skeletons were also found at some places.

The media made the tragedy Kedarnath centric calling it 'Kedarnath tragedy' as most of the people died in Kedarnath and Mandakini valley. Actually it was a pan-Uttarakhand tragedy and the destruction also happened in Himachal and Nepal. The government and media people had failed in collecting the information from other valleys. We will never know the number and names of the people who came as pilgrims and lost their lives. The large number of labourers working with pony/horse, *doli* (palanquin) and *doka* (chair) operators were largely from Nepal.

The flood was much more devastating than what the region experienced in 1894 (Birehi and Alaknanda), 1970 (Birehi, Alaknanda and Kali) and 1978 (Bhagirathi). More than 24 hydro projects (barrages, diversions, canals, tunnels and power houses) were destroyed or damaged and some of them washed away. Parts of many settlements were washed away. Not a single brick or house stone was visible at Rambara, halting point between Gaurikund and Kedarnath, where dozens of houses, hotels, restaurants were standing before the flood.

All the rivers including Tons, Yamuna, Bhagirathi, Pindar, Saryu, Ramganga East, Gori, Kali, Dhauli East and their tributaries were filled with large volume of water, silt and other floating things. Most of the agricultural fields were washed away or submerged in water. Death reports also came from Himachal and Western Nepal. The temporary bridges at many places and a hanging bridge were swept away at Jauljibi town at India-Nepal border. The dams could not play any role in regulating or controlling the floods, rather they contributed in making the floods more devastating.

6. The Story of Kashmir Floods

In September 2014 *Kashmir* region suffered disastrous floods across the state caused by torrential rainfall. The Indian state of *Jammu and Kashmir*, as well as *Azad Kashmir*, *Gilgit-Baltistan* and both *Punjabs* were affected by these floods. By 24 September 2014, nearly 277 people in India and about 280 people in Pakistan had died due to the floods. The *Jammu and Kashmir* state and adjoining areas received heavy rainfall from 2 September 2014 onwards, during last stage of *monsoon in India*.

The heavy rains triggered landslides and flooding in Kashmir and the adjoining areas of Pakistan. On 5 September the *Jhelum River* in *Srinagar* was flowing at 22.40 feet (6.83 m) which was 4.40 feet (1.34 m) above the danger mark and at 33 feet (10 m) at *Sangam* in Anantnag district above the danger mark. The discharge rate in the river was recorded as 70,000 m^3/s against the normal discharge of 25,000m^3/s. It is to be noted that Jhelum River originates from within the valley. The *Chenab River* was flowing above the danger mark by which hundreds of villages were affected in *Pakistan*. These rivers flooded into the streets causing heavy casualties and loss of property.

Several thousand villages across the state had been hit and 390 villages had been completely submerged. 2600 villages were reported to be affected in *Jammu and Kashmir*, out of which 390 villages in Kashmir valley were completely submerged. 1225 villages were partially affected and 1000 villages were affected in *Jammu Division.* Large parts of *Srinagar* were inundated and vital buildings, hospitals and roads were submerged, by the floods. The strongest post monsoon storm ever recorded in Pakistan took place on 3 September 2014 on a very low pressure system which affected parts of *Jammu and Kashmir* and North-East Pakistani districts. If we look at the chart of rains between 3 to 5 September 2014 it was between 202 mm. in Jhelum, 498 mm. in Lahore and 981 mm. in Rawalkot.

On 8 September, in many parts of Srinagar's neighbourhood, the water was about 15 feet (4.5 m.) deep, submerging entire houses. Stranded residents left their homes to move to homes in safer areas. The death toll till 10 September had crossed 190 in *Kashmir valley* and areas affected by the floods were mostly districts in South Kashmir.

Most of the city areas in Srinagar were submerged under water. Some houses and hotels were under 24 feet deep water. The loss of crops and fruits was huge. More than 50 bridges were either washed away or damaged across the state. More than 300 people lost their lives during the floods in J&K and more than one million people were affected by the floods. The most difficult thing to understand the floods in Kashmir valley was that the flood water level was so high in such a wide valley. The one cause may be heavy rains but the equally strong point is that most of the drainage points were blocked by human intervention.

7. Conclusion

It is difficult to compare the Himalayan floods with that of Kerala. The Himalaya and Western Ghats are less similar and more different. There may be some similarities in flora and fauna, crops, migration and 'money order economy'. But these are regions with different rain patterns and climatic conditions. Compared to Kerala, the Himalaya has accumulated snow and ice. Himalaya is more fragile and with more tectonic and seismic activity. It has direct international borders. Many trans-boundary rivers are coming from Tibet to India and Nepal and finally to Indus-Ganga plains.

But the ways of destruction; influx of big tourism, pressure on forests and wilderness, priority to plantation and change in land use, along with the new liberal economy, corporate domination, privatization, globalization and consumerism are common to both the regions. It is clear that conservation, equity, understanding the climatic changes and their impact on everyday life is a concern for both Kerala and the Himalaya region. Natural calamities will be there but we will be able to minimize their fury and save many human lives. Kerala and Himalayan floods gives us independent lessons for respective regions and joint lessons for both the regions.

The idea of 'Gross National Happiness' (GNH) has emerged in a small Himalayan country, Bhutan. Do we have enough courage to tilt towards it? As Gandhi has told 'The earth has enough for everyone's need, but not enough for everyone's greed', this wisdom can be at the heart of the future programmes for Western Ghats and Himalaya together.

Notes

1. This article is an edited version of the special address given at the National Seminar on Common Property and Citizens Rights: Issues of Reconstruction of Kerala, Organized by C. Achutha Menon Foundation, 25th - 26th October 2018 at Thiruvananthapuram, Kerala. Thanks to Himanshu Thakkar for the comments and suggestion on the first draft.)
2. The research was carried out by scientists from the University of Southampton based at the National Oceanography Centre Southampton (NOCS), and the Pacific Geo-science Centre, Natural Resources Canada and published in Geophysical Research Letters, a biweekly peer-reviewed scientific journal of geo-science published by the American Geophysical Union, May 2013.

Index

B

C

D

L

M

N

O

Y

Z